Springer-Verlag France S.A.R.L

Microbial Ecology and Intestinal Infections

Eugénie Bergogne-Berezin (Ed.)

Robert Ducluzeau, Marc Cerf, Gérard Corthier, Jean-Claude Rambaud, Paul Chapoy, Yannick Aujard, Gary W. Elmer

Springer-Verlag France S.A.R.L

Professeur E Bergogne-Berezin
Hôpital Bichat
Département de Microbiologie
46, rue Henri Huchard
75877 Paris Cedex 18
France

Cover illustration : Filamentous bacteria attached to a Peyer's patch of the ileum (Courtesy of R. Ducluzeau/Photo by INRA).

© Springer-Verlag France 1989

Originally published by Springer-Verlag France, Paris, 1989

ISBN 978-2-287-59510-3   ISBN 978-2-8178-0922-9 (eBook)
DOI 10.1007/978-2-8178-0922-9

2918/3917/543210 — Printed on acid-free paper.

Proceedings of the Symposium
« Microbial Ecology
and Intestinal Infections »

International Congress
for Infectious Diseases
Rio de Janeiro — April 17-21, 1988
F.-A. Waldvogel, President

Table of contents

List of authors

Y. Aujard
Hôpital Robert Debré, service de Néonatologie, 48, boulevard Serurier, 75019 Paris, France

O. Beretta
Hôpital Saint-Lazare, service de Médecine, 7 bis, rue du Faubourg Saint-Martin, 75010 Paris, France

E. Bergogne-Berezin
Hôpital Bichat, Département de Microbiologie, 46, rue Henri Huchard, 75018 Paris, France

E. Bingen
Hôpital Robert Debré, service de Bactériologie-Virologie, 48, boulevard Serurier, 75019 Paris, France

A. Bourillon
Hôpital Robert Debré, 48, boulevard Serurier, 75019 Paris, France

M. Cerf
Hôpital Louis Mourier, 178, rue des Renouillers, 92701 Colombes Cedex, France

P. Chapoy
Clinique de la Résidence du Parc, département de Gastroentérologie pédiatrique et de Nutrition, rue Gaston Berger, 13362 Marseille Cedex, France

J. Chinn
Department of Medicine, University of Washington, Seattle, WA 98195, USA

G. Corthier
INRA, Laboratoire d'Écologie Microbienne, CRJ, 78350 Jouy-en-Josas, France

R. Ducluzeau
Centre National de Recherches zootechniques, Laboratoire d'Écologie Microbienne, Domaine de Vilvert, 78350 Jouy-en-Josas, France

G.-W. Elmer
Department of Medicinal Chemistry, University of Washington, Seattle, WA 98195, USA

N. Lambert-Zechovsky
Hôpital Robert Debré, service de Bactériologie-Virologie, 48, boulevard Serurier, 75019 Paris, France

Ph. Marteau
Hôpital Saint-Lazare, service de Médecine, 7 bis, rue du Faubourg Saint-Martin, 75010 Paris, France

H. Mathieu
Hôpital Robert Debré, service de Néonatologie, 48, boulevard Serurier, 75019 Paris, France

L.-V. McFarland
Department of Epidemiology, University of Washington, Seattle, WA 98195, USA

J.-C. Rambaud
Hôpital Saint-Lazare, 7 bis, rue du Faubourg Saint-Martin, 75010 Paris, France

I. Sobhani
Hôpital Saint-Lazare, service de Médecine, 7 bis, rue du Faubourg Saint-Martin, 75010 Paris, France

P. Speelman
Department of Medicine, University of Washington, Seattle, WA 98195, USA

C.-M. Surawicz
Department of Medicine, University of Washington, Seattle, WA 98195, USA

G. Van Belle
Department of Biostatistics, University of Washington, Seattle, WA 98195, USA

Microbial ecology and intestinal infections

E Bergogne-Berezin

It has been known for a long time that the human gastrointestinal tract contains a large number of microorganisms belonging to various species and genera. Many reports have dealt with the composition of the human intestinal flora and with the isolation, identification and enumeration of hundreds of species constituting the normal flora of the gut. All major groups of microorganisms are represented, and bacteria are predominant ; protozoa of many types may be present ; yeasts are frequent hosts as well and the occurrence of bacteriophages has been demonstrated. Early studies have shown that anaerobes are the predominant bacteria ; the bacterial content of the gut may reach a number as high as 10^{14} prokaryotic cells, i.e. 10 times the number of cells in the human body. Development of microbial populations in the digestive tract commences soon after birth and a complex series of processes involved in their establishment occurs subsequently.

Many microbial-host interactions have been analyzed and many determinants for establishment of a normal gastrointestinal flora are better understood today. For instance, microbial development in the gut of mammals and in the human gastrointestinal tract depends mainly on the rate of passage of nutrients, and on the degree of acidity of the contents ; the normal stomach cavity is sterile since the normal acidity of the stomach does not constitute a suitable environment for microbial growth ; gastric bacterial overgrowth may occur in pathological conditions, in the presence of hypochlorhydria or when antacids or H_2-blockers are used in patients. Intestinal microbial colonization is mostly food-borne ; its composition results from environmental conditions and from the rate of passage of digesta. The main characteristics of the normal gastro-intestinal flora are well defined : *a)* a normal flora is made up of only certain particular organisms ; *b)* it is relatively constant over time ; *c)* there is a definite geographic localization of specific organisms ; *d)* in normal individuals the intestinal flora tends to be stable. In the concepts of microbial ecology of the gut which prevailed in the 1970s, the huge number of intestinal bacteria was considered as divided into 2 main categories ; *autochthonous* (i.e. indigenous) flora, and *allochthonous* microbes (transient bacteria) ; the former is constituted of stable populations in normal human beings, intimately associated with the mucosal epithelium ; although roughly stable, this autochthonous flora varies as a function of the segments of the gastrointestinal tract : the lower the segment, the more diversified the flora ; its equilibrium is modulated by various local or environmental factors such as oxygen tension, nutrient gradient, epithelial cell turnover, peristaltic movement and mucus flow along the mucosal surfaces.

As for the transient bacterial population, this is extremely variable ; it also originates from the environment and may colonize the epithelial surface. In fact, several situations must be considered for the transient microorganisms :

a) in normal circumstances allochthonous bacteria cannot colonize the gastrointestinal tract ; this refers to the concept of *colonization resistance* by the normal host to the implantation of « new » strains, which are unstable and eliminated rapidly from the gut. *b)* In normal human populations, *acute food-borne pathogens* can colonize temporarily the gut epithelial surface and then may invade the epithelium : this occurs with *Shigella*. In fact, many food-borne pathogens are responsible for sporadic cases or epidemics of acute diarrhea ; some of them have for long been known as allochthonous pathogenic bacteria such as *Salmonella* spp., enterotoxigenic *Escherichia coli, Vibrio* spp. More recently, other acute pathogens were recognized as being responsible for acute gastrointestinal disorders : *Yersinia* spp., *Aeromonas* spp. ; *Campylobacter* spp. ; *Bacillus cereus. c)* Of course in *debilitated human populations,* in certain geographic areas with poor conditions of hygiene and nutrition, wider and more severe epidemics of the same food-borne acute enteric infections are much more frequent : they constitute a worldwide social, health and economic problem. *d)* In developed countries, *under abnormal circumstances* in hospitals, when host defence systems are impaired, bacterial colonization may result in invasion and then subsequent infection ; this occurs in intensive-care patients in whom the digestive tract is colonized constantly from the environment by strains, transmitted mostly by hospital personnel ; *Pseudomonas, E. coli, Acinetobacter* : these transient bacteria involved in nosocomial infections are generally low-virulence species : they are more virulent in immunocompromised hosts ; but large inocula of exogenous bacteria of low virulence may also overwhelm normal defences and result in gastrointestinal disorders. In this particular situation of patients hospitalized in intensive care units, another major (though infrequent) problem is the occurrence of pseudomembranous colitis (necrotizing enterocolitis) in relation to the implantation and overgrowth of *Clostridium difficile,* a toxigenic sporulated anaerobe, generally associated with certain antibiotic therapies (amoxycillin, clindamycin, some cephalosporins with major biliary excretion).

Thus several new concepts have arisen in the microecology of the gut due to major improvement in microbiology techniques, and to the dramatic changes in environmental and therapeutic situations, particularly with the development of intensive care units and of antibiotic therapy.

First, it should be recognized that a large number of studies has led recently to a better understanding of the physiological role of the intestinal microflora, of host-bacteria interactions and of various mechanisms of gastrointestinal disorders in relation to the microbial ecology of the gut. Because of their multiplication and metabolism, the bacterial cells in intimate contact with the epithelial cells in the gastrointestinal tract play an important physiological role in human beings of all ages.

Studies with animal models have shown that the anaerobic component as well as facultative aerobes of the intestinal microflora are essential in the maintenance of equilibrium in the host-bacteria interactions. The considerable protective role played by the autochthonous flora of animal or human digestive tract has been demonstrated and this results in the concept of a « *barrier effect* » exerted by the gastrointestinal tract. This constitutes the first chapter of this book ; it is devoted to the « role of experimental microbial ecology in gastroenterology », by R Ducluzeau et al.

Association between bacteria and human intestinal cells has been well documented in various animal models and microbial adhesion to intestinal epithelial cells is considered today as a major step in bacterial colonization of the gut. In a study of « Bacterial adhesion in gastrointestinal diseases », M Cerf reports current data and personal studies on fimbrial adhesion of enterotoxigenic *Escherichia coli* in acute diarrhea and in chronic diarrheal disease. He focusses his study on biopsies taken from the jejunum and shows that true adhesion of significant numbers of bacteria may be found in some gastrectomized patients, certainly in relation to hypochlorhydria. These data confirm recent in vitro studies of adhesion of enteropathogenic *E. coli* to cultured human intestinal mucosa. It has been emphasized that the adherence of bacteria to epithelial cells involves specific structures or activities of the receptor cell, and that variations in « receptiveness » of host epithelial cells, genetically determined, are responsible for the relationship between adherence and infectivity. The latter approach is of importance in understanding bacterial virulence mechanisms and host-bacteria interactions.

Any modification in one of the constituents of the *intestinal ecosystem* may result in various disorders of the gastrointestinal tract. The administration of antibiotics, even by the parenteral route, has a number of potentially adverse effects in relation to the microbial flora : the overgrowth of autochthonous microorganisms or of yeasts may produce systemic infections in immunocompromised patients ; the overgrowth of *Clostridium difficile* and its resulting severe colitis has been mentioned above : data obtained from experimental studies in « gnotobiotic mice » are described in the chapter entitled « ecological means of protection against *C. difficile* infections in gnotobiotic mice » by G Corthier. Pseudomembranous colitis or antibiotic-associated diarrhea are observed when *C. difficile* reaches high levels of bacteria in relation to the reduction of normal microflora and of the normal « colonization resistance ».

« The pathophysiology of *Clostridium difficile*-related intestinal disease » is described by JC Rambaud et al in a chapter devoted to analysis of the mechanisms of the pathogenic role of this anaerobic bacteria in man. The possible prevention of this severe intestinal disorder is also described in this chapter.

The physiological basis for colonization resistance is generally attributed to competition of various autochthonous microorganisms for nutrients, competition for attachment sites, production of bacteriocins and production of volatile fatty acids, the latter being produced mainly by the anaerobic flora. Various disorders other than *C. difficile*-related colitis may result from intestinal bacterial overgrowth and this is analyzed in the chapter written by P Chapoy : « Bacterial overgrowth in children with severe gastrointestinal disorders ». These disorders are associated with various congenital or acquired pathologies in children such as structural defects, hemodynamic failure, immune deficiencies, protein malnutrition. This demonstrates again the protective role of the intestinal ecosystem, which is particularly vulnerable in children ; treatment and prophylaxis of bacterial overgrowth are also considered.

In the chapter entitled « Effects of antibiotherapy on microbial intestinal ecosystem in newborns and children », Y Aujard et al have analyzed the impact of antibiotic therapy using certain betalactams on the microbial equilibrium of the gut : by using « a differential quantitative analysis technique », it has been shown that the antibiotics may induce an overgrowth of ampicillin-resistant *Klebsiella pneumoniae* and *E. coli* ; a concomitant decrease of ampicillin-susceptible *Escherichia coli,* and an overgrowth of *Candida* spp. constitute a threatening situation in babies. The report of several cases of septicemia in relation to intestinal overgrowth of resistant species confirmed that antibiotic therapy may be a major cause of nosocomial septicemia in neonates. Again, a careful monitoring of intestinal microflora is of importance especially in newborns.

Another important chapter by CM Surawicz et al is dedicated to « Antibiotic associated diarrhea : risk factors and reduction of incidence by the yeast, *Saccharomyces boulardii* ». In this chapter, the role of antibiotic-associated diarrhea is underlined and its possible prevention by exogenous administration of non-pathogenic microorganisms is examined. Many years ago, several attempts were made to replace those components of the intestinal microflora that are destroyed by antibiotic therapy : early studies using *Lactobacillus* preparation, or yoghurt consumption or fermented milk containing *Lactobacillus bulgaricus* have not proven their efficacy in restoring normal equilibrium of the intestinal microflora. This was the early concept of « friendly » bacilli supposed to control bacterial overgrowth and of « substitution flora » to reestablish disrupted equilibrium of normal bacterial populations of the gut.

Based initially on empiric observation of favorable effects of absorption of a yeast, *Saccharomyces boulardii* was used in the prevention or correction of various causes of diarrhea ; today, scientific bases for its use result from a careful analysis of the mode of action ot this yeast. Lyophilized preparations of *Saccharomyces boulardii** are widely prescribed in some Western European countries, mainly as an adjuvant oral therapy to antimicrobial drugs. Early clinical studies suggested that this association had beneficial effects in preventing the occurrence of side-effects of broad spectrum antibiotic therapy related to harmful action of antibiotics on the equilirium of the normal intestinal microflora. The use of such an adjuvant to antibiotic therapy and as a preventive procedure for various intestinal disorders remained controversial for long, or at least was considered as totally useless although safe. During the last few years relevant data have been obtained from in vivo studies in animal models as reported by R Ducluzeau, or from clinical trials in patients, as underlined in most chapters of this book : the prevention or control of severe antibiotic-associated diarrhea is shown in studies by JC Rambaud, G Corthier, P Chapoy. Especially in the presence of *C. difficile* colitis, the diminished production of toxins, and in the presence of *C. albicans,* the decrease of its over-

---

* Ultralevure, Biocodex, France

growth constitute the bases for favorable clinical control by *S. boulardii* of various severe intestinal diseases. In in vitro experiments the antagonistic effects of *S. boulardii* have been confirmed against bacteria involved in diarrhea in intensive care unit (ICU) patients (personal studies) : these data confirmed the role of *S. boulardii* when used as a systematic adjuvant to antibiotic treatment in ICU patients, and in decreasing the incidence of pullulation of *Pseudomonas aeruginosa* frequently associated with diarrhea in these patients. As for the intimate mechanisms of the antagonistic effects of the yeast against some bacterial species involved in intestinal disorders, further investigations are required, though recent experiments have provided significant results, e.g. in the interactions of *S. boulardii* suspensions with enzymatic activities of brush border membranes of the intestinal mucosa. Direct interactions between *S. boulardii* and bacteria or *Candida* spp. remain to be analyzed as to their mechanisms.

Acknowledgments. It is my pleasure to thank Dr Bernasconi who made possible the International Symposium on « Microbial Ecology and Intestinal Infections » which has resulted in this book. I wish to thank also the organizers of the International Congress for Infectious Diseases who welcomed the symposium in Rio de Janeiro in April 1988, and especially Pr MB Correa-Lima of the « Clinical Medica da Universidade do Rio de Janeiro » who made personal efforts for the success of the meeting. I thank MJ Julliard for her efficient secretarial assistance and M Bornet for her excellent collaboration in in vitro studies with *Saccharomyces boulardii*.

Role of experimental microbial ecology in gastroenterology

R Ducluzeau

Newborn humans or animals are generally born without any kind of bacteria. This situation is only transitory. Bacterial development starts in the digestive tract within some hours after birth. Subsequently, man and all warm-blooded animals live, grow up and die carrying in their intestine a huge population of live bacterial cells. Thus, Luckey and Floch [26] calculated that an adult man permanently harbors about 10^{14}, hundred thousand billion live bacteria, whereas this same man's body is composed of only 10^{13} cells, 10 times less than the bacteria. The food ingested by the host and impregnated with its various digestive secretions makes it possible for these bacteria to multiply in the digestive tract of the host. The bacteria themselves also secrete numerous substances in the digestive cavities of the host and they release their intracellular components also into them when they die. For a long time the role of these bacteria in the digestive tract was neglected as microbiologists focussed their attention on pathogenic bacteria. Intestinal bacteria are gain brought into focus through a new branch of microbiology : microbial ecology.

Microbial ecology research is distinguished from medical microbiology by an important conceptual difference. It is evident that relationships between a pathogenic bacterium and the organism it attacks are highly precarious. Either the bacterium succeeds in multiplying and eliminates the host or the host succeeds in eliminating the bacterium with the help of its own defences or of the chemical weapons given to it. This is just the opposite of a relationship of the ecologic type which becomes established within an ecosystem between a group of microbial populations coexisting in equilibrium in a spatiotemporally defined region. This is the case in the digestive tract. In this enclave of external environment, limited by a living wall, hundreds of different bacterial populations keep in balance thanks to a delicate set of interactions between the different biotic and abiotic components of the ecosystem. The question may be whether these microbial populations in equilibrium constantly act in favour of the host or whether the host is permanently attacked by some of them, and in this case what can be done to improve the relations between host and bacteria. It is also certain that many diseases of the digestive tract are linked with a profound imbalance of its microbial ecosystem. The favorable or unfavorable role played by the various components of the flora of the digestive tract cannot be understood if the host cannot be completely separated from its flora. This is impossible in man and requires experimental animal models which have formed the foundation of the important development of our knowledge in microbial ecology for about 20 years.

Experimental microbial ecology tools

In order to study the relationships between the host and various components of its microflora it is essential to use completely bacteria-free animals called axenic animals. Comparing these axenic animals to animals associated with a complex flora (holoxenic) or with some known bacteria (gnotobiotic) allows us to define the role played by the microbial flora as a whole or by some of its components in the physiology of the host.

These animals must be raised in completely sterile chambers called isolators. They live on sterile feed and breathe sterile air. In France, we now have very effective isolators allowing the use of axenic animals in any biology laboratory. These chambers have also been adapted in order to maintain newborn humans germ-free (« bubbles ») or to protect adults, who have become susceptible from the microbial environment [12].

Techniques of sterile Cesarean operations or, more often, of decontamination at birth of the newborn animal or human delivered in the normal way are to place animals or human babies in germ-free conditions. Laboratory animals, such as rats and mice, are those mostly used in the axenic state but occasionally other models are preferred. Thus, the axenic piglet is highly valuable when studying the factors of establishment of the normal flora in the newborn. In fact, it is one of the few species which can easily be fed artificially from birth. Therefore, it can be used to evaluate the effect of various milk replacers (Fig. 1). The gnotobiotic bird (chicken or quail) has proved to be the only model at present allowing easy reproduction of necrotizing enterocolitis of the human newborn [28]. The pathogenic role played by toxinogenic strains of *Clostridium difficile* was simultaneously shown in the hamster affected by clindamycin treatment by two Anglo-Saxon teams and in the axenic hare in our laboratory [4, 25].

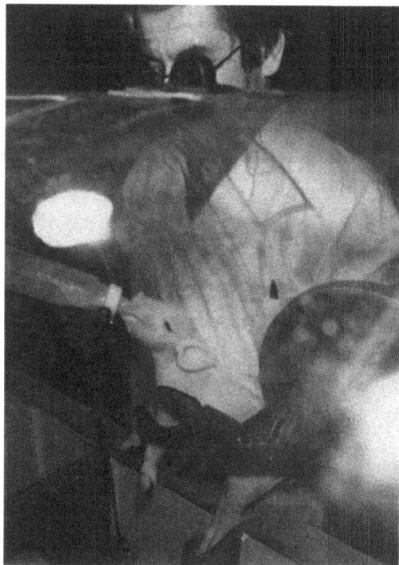

Fig. 1. Feeding of an axenic piglet in a plastic isolator

Another very useful tool in experimentation on the human gut flora was developed in our laboratory, i.e. the mouse with a human microflora. This concerns axenic mice inoculated per os with a human fecal flora. At least during the first weeks following inoculation it was noticed that the flora which became established in the mice was very similar to that of the donor and very different from that of holoxenic mice. For instance, it was possible to establish that the absorption of antibiotics by the mice with human flora resulted in effects mimicking exactly those observed in patients receiving the same antibiotics [1]. Thus, this is one of the rare cases where it is possible to experiment directly on human flora by an intermediate axenic animal.

The microbial ecologist also has to master the microbiologic techniques for quantification of the different bacterial populations intended for use. Techniques of quantitative bacteriology are tedious and still constitute one of the major brakes to microbial ecology research. They concern above all techniques of counting viable cells in dilutions of inoculated samples in series of selective culture media. Unfortunately, at present we do not have any culture medium allowing all the bacterial species of the digestive tract to grow and no selective medium is available for each bacterial population. Therefore, the cultivable bacteria of the dominant population which can be enumerated on nonselective media are well-known. On the other hand, in the subdominant flora only some groups, for which we have good selective media, are easily detected and this often leads us to give them an importance disproportionate to their real activity : this is, for instance, the case with the famous *Escherichia coli* which, despite its renown, is very subdominant in the colonic flora of the adult man. Another characteristic difficulty of the digestive tract ecosystem is due to the nature of the most numerous bacterial species : they are strictly anaerobic, i.e. they can only develop in reducing air-free media (Fig. 2). Among these strictly anaerobic bacteria some are moreover extremely sensitive to atmospheric oxygen and must constantly be kept sheltered from the air, as in the digestive tract. For this reason a kind of plastic isolator is used in which a highly reducing atmosphere is maintained by means of a gas mixture : N_2, H_2, CO_2 and a catalyst combining most of the residual oxygen to hydrogen. Inside these anaerobic chambers, called Freter's chamber after the American microbial ecologist who developed them, it is possible to work as on a laboratory bench-top, but not as easily as gloves are needed. These techniques have been improved, but we are not yet able to obtain an exhaustive « photograph » of the flora of the holoxenic digestive tract at a given moment. Therefore, flora variations can only be described indirectly through variations of indicators, i.e. selectively countable populations. This technique, which is very tedious when studying a large number of indicators, is still the most currently used in many research laboratories. However, the results are difficult to interpret as many of the data from the literature are contradictory [11, 12]. Because this technique is very complex the most obvious results concerning the impact of a nutrient or of a drug on the flora come from research on gnotobiotic animals, in which all the elements of the flora are known, rather than from holoxenic animals with a complex flora. In gnotobiotics the possibility of studying individually each bacterial population of the digestive tract allows us to state whether or not the tested factor plays a role in the equilibrium between these bacteria. Thus, potential effects of this factor on specific bacteria are revealed and it then remains to be tested, on the same bacteria, whether they are able to act in the complex flora of a live host in unprotected conditions. This approach appears to be the most suitable for understanding the mechanisms through which the dietary regimen may act upon the equilibrium of the flora.

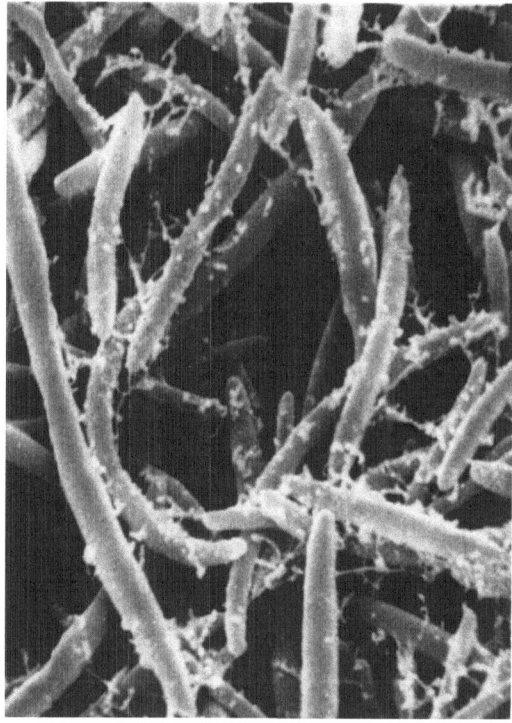

Fig. 2. Extremely oxygen-sensitive *Clostridium* in the caecum of mice

Another more recent analytic approach consists in considering the flora through its metabolic functions and in following its variations through the determination of various metabolites or the pattern of various functions. Thus, it is possible to determine the production of volatile fatty acids, bile-salt derivatives, various enzymes and gases like hydrogen or methane in the colon contents or in the feces. The pH variation or the efficiency of a barrier effect against various pathogenic microorganisms can be estimated. Naturally, this approach is not immune from criticism. Usually different bacterial groups participate in the production of one and the same substrate or one and the same activity. Thus, there is no unequivocal correspondence between a given bacterial population and a given function. Moreover, variations in metabolite quanttity can be attributed either to a quantitative variation in the bacterial populations concerned or to a modification of the metabolism of these populations without being able to settle between the two hypotheses. However, this functional approach is technically easy and perfectly suitable for estimating the impact of an exogenous factor on flora functions independently of the bacterial populations producing this effect. From this point of view the hydrogen test, which is usually called the « breath-test », was developed in man. The patient ingests an unabsorbable sugar such as lactulose and the hydrogen in the gas that he/she is breathing out is determined. The occurrence time of the hydrogen peak and the quantity of hydrogen breathed out give information on the fermentative activity of the flora, or more exactly on the activity of bacterial populations which are able to produce hydrogen from this substrate. This fraction of the flora does not play particular role, but the hydrogen diffusion rate in the tissues as well as the ease of measuring hydrogen in the respiratory gases explain why this metabolite was chosen.

Characteristics of the microbial ecosystem in man and monogastric animals

When bacterial populations in different segments of the digestive tract, from mouth to anus are examined, we are dealing not with one, but with several successive microbial ecosystems (Fig. 3).

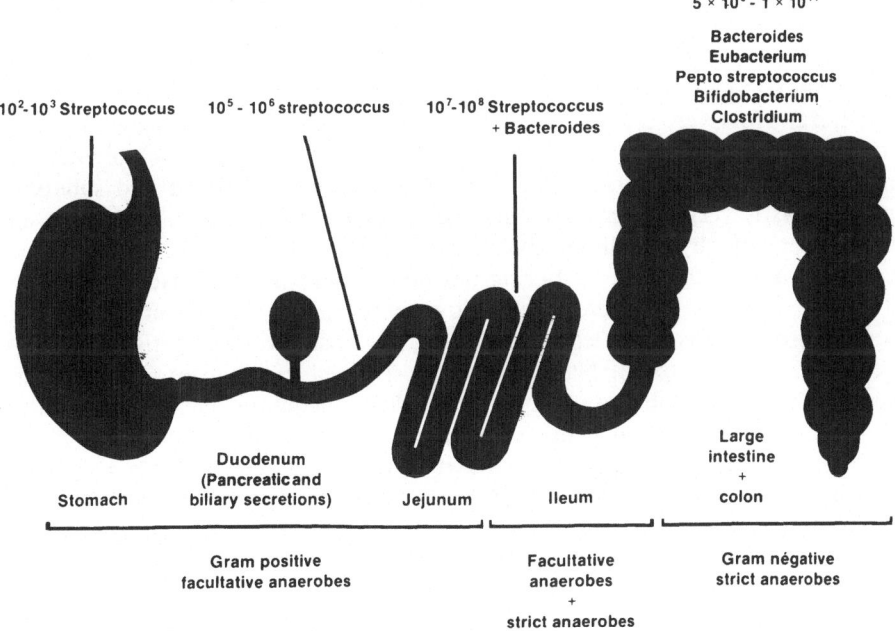

Fig. 3. Localization of the main groups of the dominant flora in the human digestive tract

The stomach always contains a more abundant flora (10^5 to 10^6 bacteria/g) when the pH is relatively high, as in the rat, and less (10^3 to 10^4 bacteria/g) when the pH is acid, as in man. This flora is primarily composed

of gram-positive facultatively anaerobic bacteria such as lactobacilli in the rat
or pig or streptococci in man. The first segments of the small intestine contain a very poor flora formed only via the transit of gastric bacteria. Closer
to the distal ileum, the flora increases in bacterial number and varieties : gram-
negative facultatively anaerobic species like enterobacteria appear next to gram-
positive species. Sometimes strictly anaerobic species are in balance with facultatively anaerobic species and their population reaches 10^9/g content. Beyond
the ileo-caecal valve and at the level of digestive stasis, in the large intestine,
caecum and colon, an explosion of strictly anaerobic bacterial species is seen.
Extremely high bacterial concentrations are reached, about 10^{10} to 2×10^{11}/g
in man. More than 190 bacterial species have been identified in the human
flora, but only some ten bacterial species cohabit at the highest population
levels. These are all strict anaerobes and most are very sensitive to contact
with atmospheric oxygen. The most abundant populations belong to the genus
Bacteroides (gram-negative) and to the gram-positive genera *Eubacterium, Bifidobacterium, Peptostreptococcus* (strictly anaerobic streptococci) and different
Clostridia [11, 14]. Facultatively anaerobic bacteria are always 100 to 1 000
times less abundant and are thus located in the sub-dominant flora (Fig. 4).
The most abundant populations in man are enterobacteria, primarily represented by the species *Escherichia coli* and streptococci. Lactobacilli frequently
found in pig, rat and mouse are scarce and always sub-dominant in the human
flora. The feces give a good indication of microbial balances in the distal gut
although some quantitative variations may take place during the passage of
intestinal contents into the rectum. Obviously, they do not reflect the intestinal flora. However, experience shows that total absence of a bacterial strain
in the feces corresponds to a total absence of this strain in all higher segments of the digestive tract. The distinction between dominant and subdominant floras is very important from a functional point of view. Indeed, it has
been experimentally established that a microbial population in the digestive tract
can only act on the host harboring it when present at levels exceeding about
5×10^7 to 5×10^8 bacteria/g. Below these values, the quantity of metabolites produced is insufficient to act on the host, however active they may
be as enzymes or toxins. Naturally, every live cell metabolizes. But the concentration of a microbial metabolite in a culture medium depends among other
things on the number of cells producing it. In a culture broth which is not
renewed 'the metabolite may increase even if only a few cells are producing
it. This is not the case with intestinal contents which are regularly renewed.
Therefore, many cells are needed to produce a large quantity of metabolites
in a short time. All bacteria playing a role in the host's physiology are thus
located in the dominant flora [14]. Bacteria in the subdominant flora, such
as facultatively anaerobic bacteria in man, are only carried by the host without
acting on it as long as their population remains repressed : the notion of a
healthy carrier is found here.

Many physiologic factors play an important role in explaining the differential flora distribution in the various compartments of the digestive tract.
In the stomach the principal factor is certainly free acidity. Many of the bacteria ingested with the dietary bolus are sorted out at this level, but the filtration is never perfect and it can be noticed that it is always possible to intro-

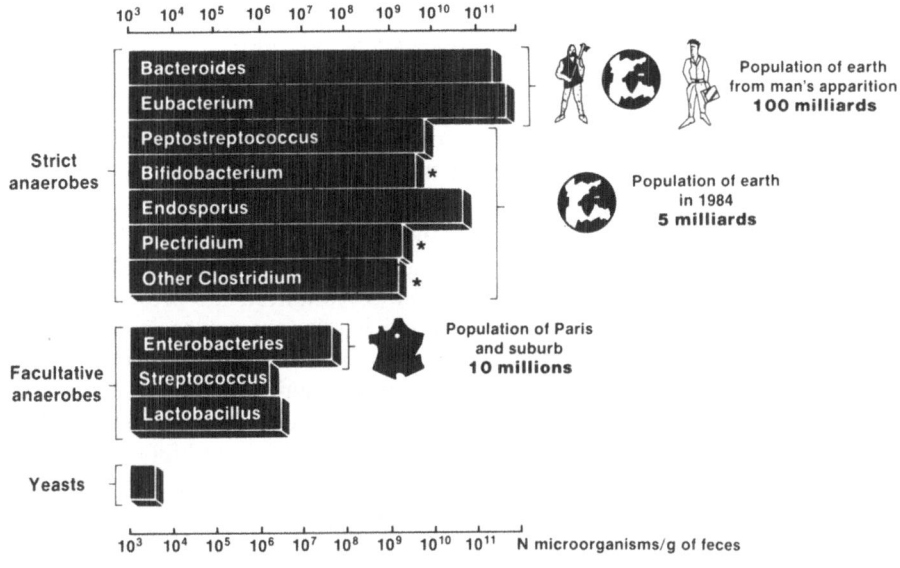

Fig. 4. Example of normal bacterial profile of human fecal flora

duce a strain in the entire digestive tract of an axenic animal by oral inoculation. However, in many pathologic conditions like hypochlorhydria or after gastrectomy a considerable increase in the flora of the stomach is observed (up to 10^9 or 10^{10} bacteria/g) and consequently in that of the proximal small intestine. Pancreatic and biliary secretions are supposed to play an important role in reducing the flora in the proximal small intestine. Actually, experimental models (rats with a controlled flora whose secretions are diverted into the bladder by a catheter) did not show large differences from the controls [30]. On the other hand, intestinal peristalsis is certainly the essential factor responsible for the elimination of intestinal bacteria. If the transit of the dietary bolus is normal, the residence time of bacteria in the intestine is shorter than the time needed for their multiplication and therefore they do not have time enough to proliferate. As soon as the transit rate slows down, as in the distal ileum, and the stasis regions of the caecum and colon are reached, the microbes divide and their total number increases [6]. However, in this distal part of the digestive tract bacteria in equilibrium divide only little : considering that the contents of the large intestine are almost emptied every day and that the highest level of the bacterial population remains almost unchanged, bacteria of the dominant flora are hardly able to divide more than once or twice a day. Any pathologic manifestation capable of diminishing or stopping peristalsis of the small intestine immediatly results in the establishment of a flora qualitatively and quantitatively comparable to that of the caecum in the segment where the slowing occurs : this is the case with blind loops, diverticulitis ans stenosis. Strictly anaerobic bacterial proliferation in the small intestine is always extremely unfavorable to the host. Actually, these bacteria will destroy or modify metabolites and nutrients normally absorbed or reabsorbed before they come into contact with the flora of the large intestine. The most striking example of these disorders is that of lethal steatorrhea, which can be induced in the rat carrying a blind loop in the jejunum. The anaerobic bacteria of the loop deconjugate the bile salts and the animal becomes totally unable to absorb fats, which are all found in the feces [31].

Thus, the so-called normal flora of the colon is only tolerated by the host when it is strictly limited to the distal compartments of the digestive tract.

There is a bacterial factor which is able to condition the presence of bacteria at various levels of the digestive tract, i.e. adhesion to the mucosa. Some bacteria have diversified antigenic structures on their surface, now referred generically to adhesins, which allow them to recognize suitable structures on the enterocytes of the mucosa [32]. Hence such bacteria are able to adhere to the wall of the small intestine in particular, and to multiply there while resisting to peristalsis. Once again, it was shown in the gnotobiotic piglet that the mere fact of proliferating at a level which normally contains no bacteria made these microorganisms detrimental to the host [15, 16] : adhesion to the wall of the small intestine is not a normal ecologic factor.

In the large intestine other bacteria adhere to the mucosa by a totally different mechanism : a chemotaxis attracting bacteria towards mucus and leading them to adhere to the mucosa of the large intestine. Adhesion to dietary particles present in the digestive contents has also been observed. These phenomena unlike the above are equilibrium factors of the ecosystem.

According to some authors any bacterium of the autochthonous flora of an animal should be able to adhere to the mucosa by any mechanism but this point is still much debated.

Stability of the equilibrium of the flora in the digestive tract

The stability of the flora equilibrium as a function of time and of exogenous factors, of which the most important remains the diet composition, is still much debated.

The few comprehensive comparative studies carried out in the axenic man using quantitative differential methods of analysis conclude in favor of an astonishing stability of the fecal flora. The most famous of these studies was performed in the USA, in the « Laboratory of Bacteriology of Anaerobes of the Medical Institute of Wordsworth ». For almost 10 years, the scientists of this Institute analysed the fecal flora of some 150 people receiving various diets. The purpose was to find a correlation between certain bacterial populations and certain diets in order to show the role possibly played by these populations in the etiology of colon cancer. The observed subjects either received a strictly vegetarian diet or a vegetarian diet with a little meat or traditional Japanese or American standard Western food. Finally, whatever the diet, no significant difference could be shown between fecal floras although the incidence of colon cancer varied according to groups [19].

In an experiment carried out in our laboratory we studied the effect of wheat-bran fibers on the flora of mice inoculated with human fecal floras from donors already receiving fibers or not. Although this was a situation where any supply of new exogenic microorganisms to the initial flora was impossible, no significant differences could be noticed between animals with or wihtout wheat-bran diets [10].

Does that mean that the diet is unable to act on the equilibrium of the flora ? Certainly not, because very many examples show the opposite in experimental models constituted by gnotobiotic animals. Only 3 examples will be given here. In the first case it was noticed that in animals fed with a semi-synthetic diet, S, it was impossible to implant a strain of *Clostridium* which was implanted at a high level in animals fed with a commercial diet, C. In this case the inhibitory effect of diet S was due to a very particular epsilon-dipeptide derived from the heated casein added to the diet. This dipeptide, which was not absorbed because of its structure, was capable of chelating the copper contained in the diet and the mixed dipeptide-copper molecule then exerted a very strong inhibitory effect on the target strain [5].

The second example once gain involves a *Clostridium* capable of becoming established in the digestive tract of axenic mice in the presence of a diet, P, called permissive and unable to become established in the presence of another diet, N P, called non-permissive. If the animals were first fed with the permissive diet, P, the strain became established but, after 2 or 3 weeks, if diet P was replaced by the non-permissive diet, N P, it was noticed that the *Clostridium* strain was not eliminated but persisted indefinitely. What is shown is in fact that the very large population of the strain present from the use of diet N P was capable of modifying it by eliminating the inhibitory factor. That is what we called the remanence phenomenon [13].

The last example involves a toxinogenic strain of *Clostridium difficile*. When inoculated into mice fed with a commercial diet, this strain became established and killed 100 % of the animals within 48 h. When inoculated into mice fed with a semi-synthetic diet, it also became established at the same level as the former one, but one animal died. Under these conditions very low quantities of toxins, in particular enterotoxins, were recovered in the digestive tract. It seems that the nature of the dietary protein source plays an important role in « in vivo » toxin production [27].

These examples clearly show that the nature of the diet is capable of acting upon bacterial implantation, multiplication and metabolic activity in the flora of the digestive tract.

Barrier effect of the gastrointestinal microflora

Comparison of axenic and holoxenic animals shows that the flora of the digestive tract plays a considerable role. Table 1 gives an account of the main fields of host physiology influenced by the presence of bacteria in the digestive tract. However, emphasis will more especially be laid on an essential function of the gut flora, i.e. its role as a barrier towards the microbial environment.

Table 1. Main physiologic functions exerted by the microbial flora of the digestive tract

. Modifications of intestinal contents : pH, EH, production or destruction of metabolites

. Anatomic modifications of the digestive tract : caecal volume (in some animals), structure of the intestinal wall (in all animals studied)

. Modifications of digestive physiology : transit of gastric and intestinal contents, enterocyte renewal rate, absorption of abiotic components

. Immune system modifications : increase in number of IgA plasmocytes in the intestine, establishment of immuno-tolerance

. Protection against invasion of the intestine by exogenous bacteria : barrier effects, modulation of toxin production in the intestine

Man and holoxenic animals live in an environment rich in all kinds of microorganisms which their gut flora does not, however, faithfully reflect. Licking a stamp, swallowing a badly cleaned salad leaf, opening the mouth in the subway, all represent daily opportunities for bacteria to enter the digestive tract. Some of these bacteria are potentially capable of causing an infectious disease. But we are not ill each time we ingest such bacteria. Likewise, ingesting a yogurt or a portion of sauerkraut corresponds to a massive uptake of lactic bacteria in the digestive tract without any colonization of the latter by these bacteria. The surprising resistance of a holoxenic host gut to colonization by daily ingested exogenous bacteria is due to « barrier effects » exerted by the « autochthonous » flora of the digestive tract normally associated with the host. All these exogenous strains, which have no chance of persisting in the digestive tract of a holoxenic animal settle without any problem in the axenic animal.

When comparing the transit of a passive marker (spores of a strain of strictly thermophilic *Bacillus*) and of an exogenous bacterium [7] through the digestive tract of an animal carrying a barrier flora, a parallelism between the two elimination curves is usually observed. It may be inferred from this that the barrier flora exerts a bacteriostatic effect on the target bacterium, which is then eliminated passively by peristalsis of the digestive tract. Some bacteria sometimes disappear faster from the digestive tract than the transit marker, which means that they are partly destroyed during transit. On the other hand, other bacteria subsist in small amounts after disappearance of the marker, which means that they tend to multiply at a low rate which is, however, high enough to compensate for the emptying due to peristalsis : here we find again « healthy carriers » of a microorganism. In this case the microbiologic barrier is called permissive, whereas it is called drastic when it succeeds in totally eliminating exogenous bacteria from the digestive tract (Fig. 5).

The use of gnotobiotic animal models carrying simplified and well-known microbial floras in various studies have made it possible to determine the main characteristics of the barrier effect.

First, it may be stated that the barrier effect against a viable bacterium

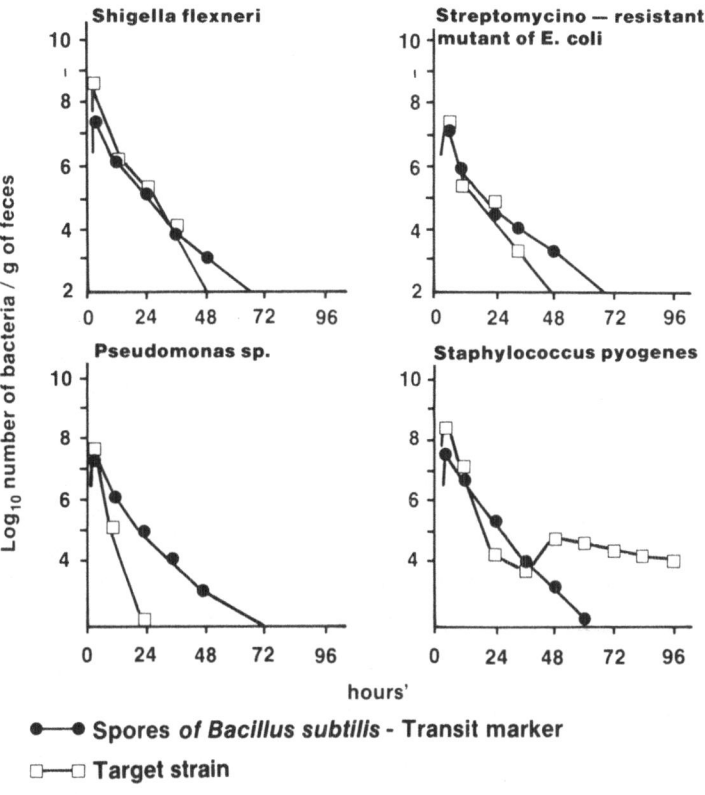

●—● Spores of Bacillus subtilis - Transit marker

□—□ Target strain

Fig. 5. Barrier effect exerted by flora of conventional mice against 4 exogenous bacterial strains. *1 Shigella flexneri* is passively eliminated, as a transit marker; *2* a streptomycin-resistant mutant of *E.coli* strain present in the flora of the same holoxenic mice, is passively eliminated with the same kinetics as an exogenous strain ; *3 Pseudomonas* sp. is partially destroyed during transit in the digestive tract ; *4 Staphylococcus pyogenes* is first passively eliminated, then persists in the subdominant flora : this is an example of a healthy carrier

is due do the action of specialized bacteria present in the dominant flora. It has too often been declared that « space is occupied by the first arrivals » in the digestive tract. This is not the case, as shown by the following experiment : a human newborn delivered by axenic Cesarian operation, was inoculated with a strain of *Lactobacillus casei* listed in the pharmacopea. This strain became established at a high level in the axenic baby and lowered the pH of the feces from 7 to 5. After having verified that the baby was without any immunologic defect it was decided to remove it from the isolator. To that end, the baby was given a flora of a normal human newborn including 4 strains of the following species : *Escherichia coli, Streptococcus* sp., *Bifidobacterium bifidum, Bacteroides* sp. It was then noticed that « the first occupant », *Lactobacillus,* was immediately eliminated and replaced by the population of *E. coli* in the dominant flora, and it was shown that this *E. coli* strain alone was able to exert a strong antagonistic effect against *Lactobacillus* [23]. Thus, strains becoming established in the dominant flora are not the first ones arrived, but those capable of driving the others out of their biotope.

When attempting to isolate bacteria exerting a barrier effect against a particular target strain in the complex microflora of a holoxenic animal we have to cope with serious technical difficulties. Accordingly, only a few simplified barriers are known. In our laboratory, we succeeded in isolating 3 strains exerting a drastic and curative barrier effect against a target strain of *Clostridium perfringens* [33] (Fig. 6). First, it was noted that the barrier was less efficient when a strain was taken out of the ecosystem. The two-strain model was less efficient and each strain taken individually was completely inefficient. Therefore, there was necessarily a synergy between these bacteria. The barrier was efficient against all the strains of *C. perfringens,* type A, tested by us. It was only partly efficient against *C. perfringens,* type C, and *C. difficile.* On the other hand it was active against an enterobacterium like *S. flexneri* which was unconnected with the genus *Clostridium.* Hence, the spectrum of action of a simplified barrier is relatively narrow and not related to the taxonomic position of target strains. Environmental factors may modulate the activity of the barrier effect. One of the most important factors is the diet. The barrier against *C. perfringens* which has just been described was fully active when gnotobiotic mice received a diet sterilized by irradiation. If the diet was sterilized by autoclaving it was observed that the barrier effect became very low, whereas the population level of the three strains dit not change. On returning to the

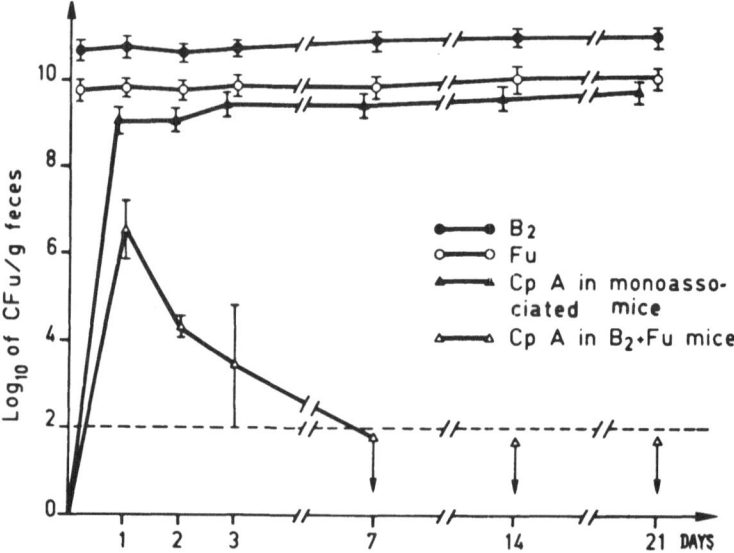

Fig. 6. Establishment of CpA in the digestive tract of axenic mice and antagonistic effect against CpA exerted by strains B₂ and Fu previously established in the digestive tract of gnotobiotic mice. The simultaneous presence of the two strictly anaerobic barrier strains B₂ (Bacteroides sp.) and Fu (Fusobacterium sp.) is necessary to obtain complete elimination of the target strain CpA (*Clostridium perfringens* type A)

irradiated diet everything went back to normal. There are also specific host factors modulating the barrier effect. The simplified barrier against *C. perfringens* was drastic in the gnotobiotic mouse, but only permissive in the gnotobiotic rat receiving the same diet, even though the population level of the three barrier strains was identical in the two types of animals. The age of the animals is involved : the barrier exerted against *E. coli* by a complex flora from mice was low in young animals (2 to 3 months) but much more efficient in older animals [9]. Finally, other still unknown factors may result in unpredictable barrier ruptures, the source of digestive disorders, if a pathogenic agent is present in the flora at the same time. Thus, a human fecal flora associated with axenic mice exerted a drastic barrier against a strain of *C. difficile* for 66 days. After that time, and without any detectable environmental modification, the fecal population of *C. difficile* suddenly reached 5×10^8 bacteria/g and remained at that level [29]. Hence, the development of digestive disorders seems to be the result of spontaneous or induced disappearance of the barrier effect and the simultaneous presence of a pathogenic bacterial agent in the ecosystem.

The mechanism of these barrier effects has still not been well established. However, some preliminary data are available.

First, there is an « intraspecific » barrier effect, i.e. an interaction between 2 strains belonging to the same bacterial species. For instance, some *E. coli* strains exert a preventive effect against other strains of *E. coli* [18]. This mechanism has not yet been clarified. It could have been assumed that an antibiotic active against the target strain was produced by the inhibitory strain, for instance a colicin. But in these trials the inhibitory strain was sensitive in vitro to colicin produced by the target strain ; therefore this mechanism does probably not involve the production of colicin.

In the case of heterospecific barriers, Freter put forward the theory of an antagonistic effect linked to competition for a growth factor essential to the target again. Studying this phenomenon in the case of a barrier exerted by a complex flora against *E. coli* he came to the following assumption. Barrier bacteria might synthetize a metabolite (probably hydrogen sulphide) which, under the conditions of pH and oxido-reduction potential of the digestive tract, might prevent *E. coli* cells from using their energizing substrate, glucose [20, 21].

In the case of the simplified barrier against *C. perfringens* we have shown that this theory is not valid. It is also impossible to find an antagonistic substance on the supernatant of caecal contents. The barrier effect seems, however, to be due to the secretion of an antagonistic metabolite in very low quantities, or of short life, which is only active when the target bacteria are almost in contact with the barrier bacteria [33].

When the barrier effects of the flora are exerted against environmental pathogenic bacteria having penetrated the ecosystem, they can be considered as a very important element of the host's defence system. Owing to these barrier effects the evolution of man and animals has been possible in an environment often very rich in pathogenic microorganisms. Moreover, it has been established that disturbing the equilibrium of the microbial ecosystem by ingestion of antibiotics allows some pathogenic bacteria to pullulate and cause disorders wich may be dramatic. It was thus discovered that ingestion of clindamycin destroys the barriers maintaining *Clostridium difficile* at very low levels in the digestive tract of very many people. *C. difficile* is then allowed to multiply and secrete its dangerous toxins in quantities large enough to kill the patient [25].

When the barrier effects are exerted against harmless environmental microorganisms they can be compared to a homeostatic effect responsible for the stability of the microbial ecosystem.

How to modify the gastrointestinal microflora in order to prevent or cure disorders of the ecosystem

Possibilities of implantation of selected microorganisms

For a very long time it was assumed that microorganisms capable of playing a useful role could be « implanted » into the digestive tract simply by ingesting huge quantities of these live microorganisms. The pharmacopea contains many preparations supposed to « rebalance » the flora : *Bacillus subtilis*, *Escherichia coli*, yeasts. This notion of « reimplantation » also explains why yogurt is often prescribed together with antibiotics. But of course the barrier effects which we described are also exerted against these deliberately ingested microorganisms, and it has now been shown that none of these bacteria or yeasts

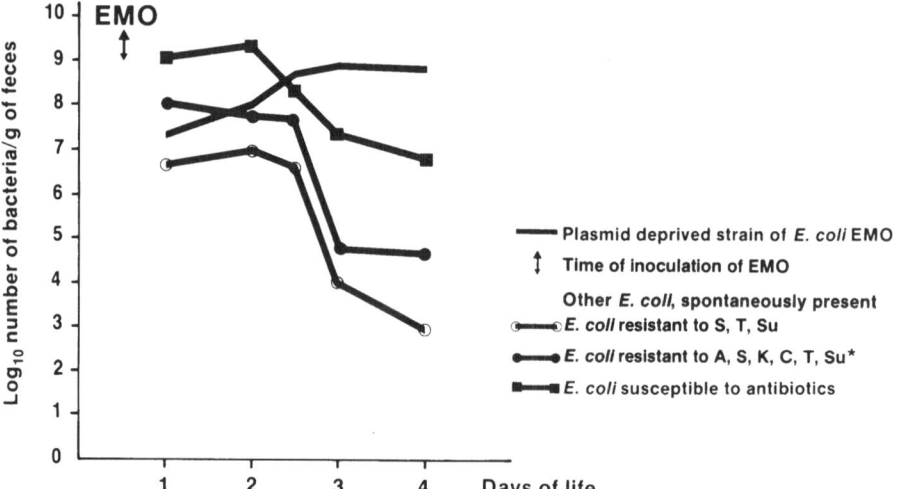

* A : ampicillin ; S : streptomycin ; K : kanamycin ; C : chloramphenicol ; T : tetracycline ;
Su : sulfonamides.

Fig. 7. Immediate postnatal inoculation of a plasmid-deprived strain of *Escherichia coli* in the digestive tract of a human new-born. The EMO human *E.coli* strain, inoculated at birth, multiplies and exerts a barrier effect against 3 other strains of *E.coli* acquired from the environment. Two of these strains bear plasmids encoding for resistance to various antibiotics

are able to multiply and durably subsist in the dominant flora of a subject already colonized by a complex autochthonous flora. Therefore, it is impossible to modify the microbial ecosystem of the digestive tract (even when disturbed by antibiotics) by the mere ingestion of live microorganisms. There is only one exception to this rule : the moment of birth when the digestive tract is not yet colonized by any microbe. A study conducted by Y Duval in the human newborn illustrates this possibility. The first bacteria to become established in the digestive tract of the newborn within 24 h after birth always belong to the species *E. coli*. However, among these *E. coli* strains some are inoffensive but others are undesirable as they have pathogenic factors (factors of adhesion to the mucosa, toxins, etc.) and antibioresistant factors often coded on extrachromosomal elements, the plasmids. A strain of human *E. coli* without plasmids and pathogenic factors was prepared and tested on animal models in order to confirm its ecologic advantage over the majority of plasmid carrier *E. coli* strains. This strain was inoculated at birth in 22 infants in a hospital. In 86 % of the cases, this strain became established in the dominant flora and eliminated simultaneously and spontaneously established strains of *E. coli,* characterized by multiresistance to antibiotics and potentially pathogenic [17] (Fig. 7).

Activity of viable bacteria during their transit in the digestive tract

The mechanism of action of biologic barrier effects is by bacteriostasis : barrier bacteria prevent target bacteria of the exogenous inoculum from growing, but they do not kill them. Then these target bacteria are eliminated by intestinal peristalsis. Therefore it is possible to make very many live bacteria pass through the digestive tract, metabolizing the substrate during their transit : microorganisms used in this way are called « probiotics ».

Two sorts of preparations used empirically for decades have shown experimentally the efficiency of viable microorganisms during gastrointestinal transit. The yeast *Saccharomyces boulardii* is used in pharmacology (Ultralevure). In gnotobiotic mice inoculated with a pathogenic yeast, *Candida albicans,* it was shown that the transit of a very large number of viable cells of *S. boulardii* resulted in a marked reduction of the *C. albicans* population in the digestive tract [8] (Fig. 8).

In another experiment, axenic mice were inoculated with a toxinogenic strain of *Clostridium difficile* which killed them within 48 h. If a very large number of *S. boulardii* was allowed to transit in the digestive tract most of the animals survived. In all the surviving animals it was observed that the population of *C. difficile* dit not change, but the quantities of cytotoxins and enterotoxins present in the digestive tract were much reduced. Again, the use of selected *S. boulardii* no longer made it possible to observe this modulatory effect on toxin production [3] (Fig. 9).

Another « probiotic », yogurt, is widely used all over the world as food. Recent studies [24] showed that patients with lactase deficiency absorb lactose of yogurt much better than that of milk. The breath test shows that only very little lactose from yogurt comes into contact with bacteria capable of fermen-

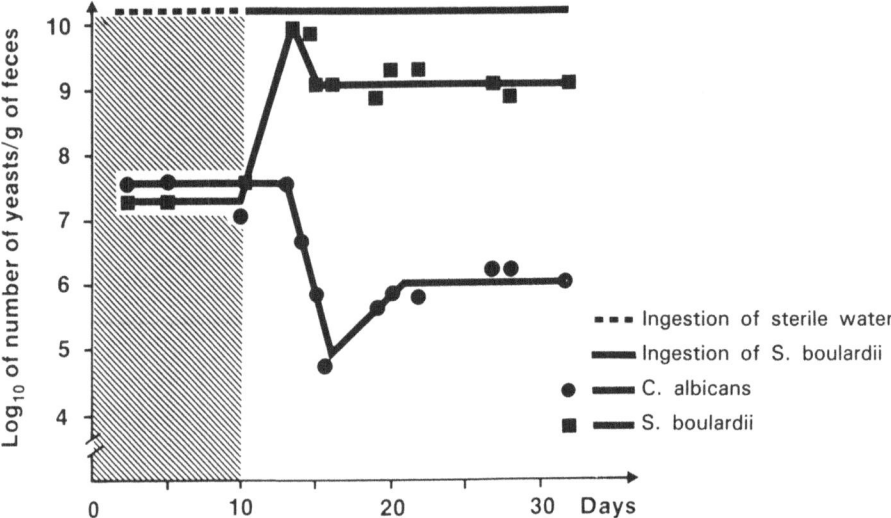

Fig. 8. Interaction between *Saccharomyces boulardii* and *Candida albicans* in the feces of gnotobiotic mice. Axenic mice were first orally inoculated at J_0 with *C.albicans* and *S.boulardii* and then received sterile water ad lib. At J_{10} they received as a fluid source a suspension containing 5×10^9 cells of *S.boulardii* per ml. During this treatment, the number of *C.albicans* decreased 50 to 100 fold

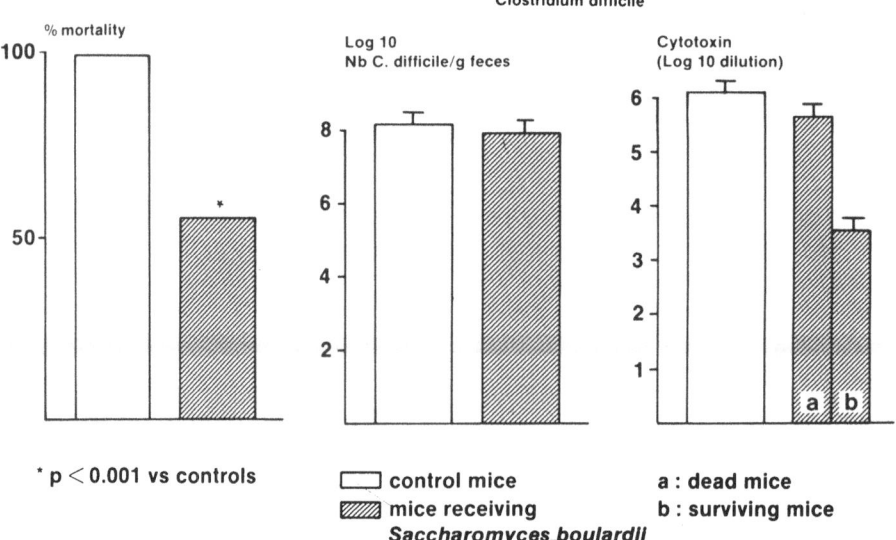

Fig. 9. Mortality level, fecal population of *Clostridium difficile* and cytotoxin titer in gnotobiotic mice some of which received a suspension of 10^9/ml *Saccharomyces boulardii*. Treated animals receiveid 5×10^{10} viable cells of *Saccharomyces boulardii* per ml of drinking water beginning at 4 days before inoculation of *C.difficile*. The level of fecal population of *C.difficile* is not modified by this treatment. However 50 % of the treated animals survived and their cecal level of cytotoxin was reduced by a factor of nearly 1000 fold

ting it in the large intestine. According to the authors, the bacteria of the yogurt continue to hydrolyse lactose in the intestine during their transit. Other authors [2] showed moreover a stimulation of lactase of the gut mucosa by transiting live lactic bacteria (Fig. 10).

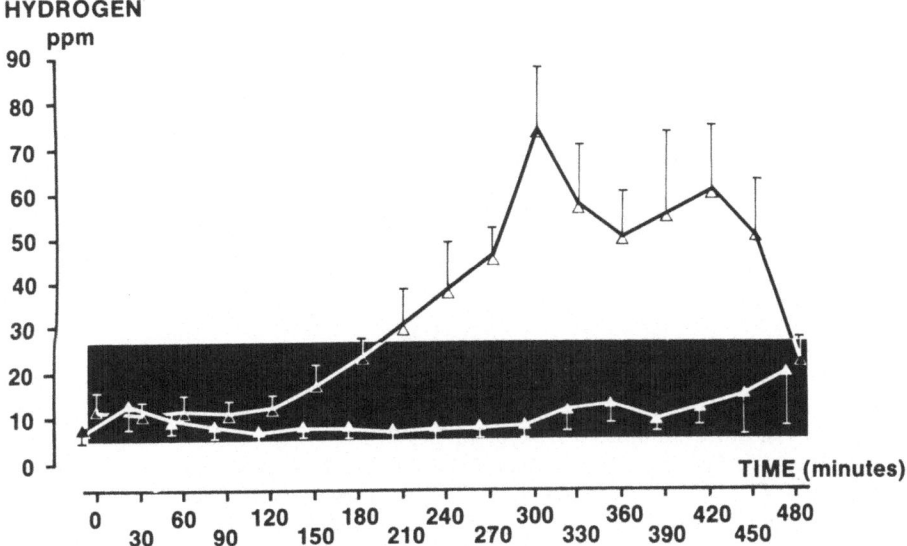

Fig. 10. Amount of breath hydrogen expelled after ingestion of yogurt (▲), and thermised yogurt (△) in lactase deficient subjects. The amount of hydrogen after ingestion of living yogurt is very low compared to the amount expired after ingestion of pasteurized yogurt, suggesting the possibility of hydrolysis of lactose by living bacteria during their transit in the intestine

Goldin and Gorbach [22] showed that ingestion by humans of milk fermented by *Lactobacillus acidophilus* resulted in a reduction of several enzymatic activities exerted by the microbial fecal flora, i.e. ß-glucuronidase, nitroreductase, azoreductase. This activity is considered as favorable because these enzymes may be involved in the genesis of carcinogenic products.

Therefore, the role of *live* microbial cells *present in very high numbers* during their transit through the digestive tract has been proven objectively in some cases. In the short term, the use of probiotics is certainly more interesting in practice than the implantation of strains in the dominant flora of the subject. However, many studies have still to be carried out to determine the potentialities of these two methods.

Conclusion

Much research remains to be done to clarify the multiple relationships throughout life, between the host, its intestinal microbes and its environment. Studies of gnotobiotic models give us fragmentary, but often unexpected data on various functions of the intestinal flora, its significance for the health of the host and the modulation of these functions by the food of the host. First, they have the advantage of suggesting an experimental approach which will put preconceived ideas in their right place, of which the majority comes from in vitro experiments and are illegitimately extrapolated to in vivo conditions. Second, this approach will allow improvement of the quality of life by preventing the dysfunction of intestinal microbes. Data already available stress the necessity of preserving to the maximum the equilibrium of the flora whenever it is possible, and also raise hopes of approaches to ecologic techniques aiming at providing the host with the most suitable ecosystem.

References

1. Andremont A, Raibaud C, Tancrede C, Duval-Iflah Y, Ducluzeau R (1985) The use of germ-free mice associated with human fecal flora as an animal model to study enteric bacterial interactions. In : Takeda Y, Miwatini T (Eds) Bacterial Diarrheal Diseases. KTK Scientific Publishers, Tokyo, pp 219-228
2. Besnier MO, Bourlioux P, Fourniat J, Ducluzeau R, Aumaitre A (1983) Influence de l'ingestion de yogourt sur l'activité lactasique intestinale chez des souris axéniques ou holoxéniques. Ann Microbiol (Inst Pasteur) 134 A : 219-230
3. Corthier C, Dubos F, Ducluzeau R (1986) Prevention of *Clostridium difficile*-induced mortality in gnotobiotic mice by *Saccharomyces boulardii*. Can J Microbiol 32 : 894-896
4. Dubos F, Martinet L, Dabard J, Ducluzeau R (1984) Immediate postnatal inoculation of a microbial barrier to prevent neonatal diarrhea induced by *Clostridium difficile* in young conventional and gnotobiotic hares. Am J Vet Res 45 : 1242-1244
5. Dubos F, Pelissier JP, Andrieux C, Ducluzeau R, Raibaud P (1985) Inhibitory effect of a copper-dipeptide complex on the establishment of a *Clostridium perenne* strain in the intestinal tract of gnotobiotic mice. Appl Environ Microbiol 50 : 1258-1261
6. Ducluzeau R (1981) La flore microbienne de l'intestin grêle et les conséquences d'une stase intestinale. XIe J Nat Néonat, Paris 1 : 98-102
7. Ducluzeau R, Bellier M, Raibaud P (1970) Transit digestif de divers inoculums bactériens introduits per os chez des souris axéniques et holoxéniques (conventionnelles) : effet antagoniste de la microflore du tractus gastro-intestinal. Zentralbl Bakteriol Parasitenkdl Infectionskr Hyg [B] 213 : 533-548
8. Ducluzeau R, Bensaada M (1982) Effet comparé de l'administration unique ou en continu de *Saccharomyces boulardii* sur l'établissement de diverses souches de *Candida* dans le tractus digestif de souris gnotoxéniques. Ann Microbiol (Inst Pasteur) 133 B : 491-501
9. Ducluzeau R, Ladire M, Duval-Iflah Y, Huet S (1981) Influence du vieillissement de l'hôte sur l'effet de barrière permissif à l'égard d'une souche d'entérobactérie, exercé par une flore bactérienne complexe dans le tube digestif de souris maintenues en isolateur. Ann Microbiol (Inst Pasteur) 132 B : 91-100
10. Ducluzeau R, Ladire M, Raibaud P (1984) Effet de l'ingestion de son de blé sur la flore microbienne fécale de donneurs humains et de souris gnotoxéniques receveuses et sur les effets de barrière exercés par ces flores à l'égard de divers microorganismes potentiellement pathogènes. Ann Microbiol (Inst Pasteur) 135 A : 303-318
11. Ducluzeau R, Raibaud P (1976) La compétition bactérienne dans le tube digestif. La Recherche 7 : 270-272

12. Ducluzeau R, Raibaud P (1979) Écologie microbienne du tube digestif. Actualités Scientifiques de l'INRA, Masson, Paris

13. Ducluzeau R, Raibaud P, Dubos F, Clara A, Lhuillery C (1981) Remanent effect of some dietary regimens on the establishment of two *Clostridium* strains in the digestive tract of gnotobiotic mice. Am J Clin Nutr 34 : 520-526

14. Ducluzeau R, Raibaud P, Hudault S, Nicolas JL (1980) Rôle des bactéries anaérobies strictes dans les effets de barrière exercés par la flore du tube digestif. Les anaérobies : microbiologie-pathologie. Masson, Paris, pp 86-95

15. Duval-Iflah Y, Chappuis JP (1984) Entéropathogénicité de différentes souches de *Escherichia coli* chez le porcelet gnotoxénique. Les colloques de l'INSERM. La diarrhée du jeune. INSERM 121 : 167-176

16. Duval-Iflah Y, Chappuis JP (1984) Influence of plasmids on the colonization of the intestine by strains of *Escherichia coli* in gnotobiotic and conventional animals. In : Klug MJ, Reddy CA (Eds) Current perspectives in microbial ecology. Gastrointest Microecol, pp 264-272

17. Duval-Iflah Y, Ouriet MF, Moreau C, Daniel N, Gabilan JC, Raibaud P (1982) Implantation précoce d'une souche de *Escherichia coli* dans l'intestin de nouveau-nés humains : effet de barrière vis-à-vis de souches de *E. coli* antibiorésistantes. Ann Microbiol (Inst Pasteur) 133 A : 393-408

18. Duval-Iflah Y, Raibaud P, Rousseau M (1981) Antagonisms among isogenic strains of *Escherichia coli* in the digestive tracts of gnotobiotic mice. Infect Immun 34 : 957-969

19. Finegold SM, Sutter V (1978) Fecal flora in different populations with special reference to diet. Am J Clin Nutr 31 : 116-122

20. Freter R, Brickner H, Botney M, Cleven D, Aranki A (1983) Mechanisms that control bacterial populations in continuous-flow culture models of mouse large intestinal flora. Infect Immun 39 : 676-685

21. Freter R, Brickner H, Fekete J, Vickerman MM, Carey FE (1983) Survival and implantation of *Escherichia coli* in the intestinal tract. Infect Immun 39 : 686-703

22. Goldin MR, Gorbach SL (1985) The effect of milk and *Lactobacillus* feeding on human intestinal bacterial enzyme activity. Am J Clin Nutr 39 : 756-761

23. Hudault S, Ducluzeau R, Dubos F, Raibaud P, Ghnassia JC, Griscelli C (1976) Élimination du tube digestif d'un enfant « gnotoxénique » d'une souche de *Lactobacillus casei* issue d'une préparation commerciale : démonstration chez des souris « gnotoxéniques » du rôle antagoniste d'une souche de *Escherichia coli* d'origine humaine. Ann Microbiol (Inst Pasteur) 127 B : 75-82

24. Kolars JC, Lewitt MD, Aouji M, Savoiano DA (1984) Yogurt, an autodigesting source of lactose. N Engl J Med 310 : 1-3

25. Larson HE, Price AB (1977) Pseudomembranous colitis : presence of a clostridial toxin. Lancet II : 1312-1314

26. Luckey TD, Floch MH (1972) Introduction to intestinal microecology. Am J Clin Nutr 25 : 1291-1295

27. Mahe S, Corthier G, Dubos F (1987) Effect of various diets on toxin production by two strains of *Clostridium difficile* in gnotobiotic mice. Infect Immun 55 : 1801-1805

28. Popoff MR, Szylit O, Ravisse P, Dabard J, Ohayon H (1985) Experimental cecitis in gnotoxenic chickens monoassociated with *Clostridium butyricum* strains isolated from patients with neonatal necrotizing enterocolitis. Infect Immun 45 : 697-703

29. Raibaud P, Ducluzeau R, Dubos F, Hudault S, Bewa H, Muller MC (1980) Implantation of bacteria from the digestive tract of man and various animals into gnotobiotic mice. Am J Clin Nutr 33 : 2440-2447

30. Sacquet E, Raibaud P, Garnier H (1971) Étude comparée de la microflore de l'estomac, de l'intestin grêle et du caecum du rat « holoxénique » (conventionnel) et de ses modifications à la suite de diverses interventions chirurgicales : anse aveugle jéjunale, déviations biliaires. Ann Inst Pasteur 120 : 501-524

31. Sacquet E, Garnier H, Raibaud P, Eyssen H (1968) Étiologie bactérienne de la stéatorrhée observée chez le rat porteur d'un cul-de-sac intestinal. Déconjugaison de l'acide taurocholique. Cr Acad Sci 267 : 2238-2240

32. Tannock GW, Szylit O, Duval Y, Rainaud P (1982) Colonization of tissue surfaces in the gastrointestinal tract of gnotobiotic animals by *Lactobacillus* strains. Can J Microbiol 28 : 1196-1198

33. Yurdusev N, Nicolas JL, Ladire M, Ducluzeau R, Raibaud P (1987) Antagonistic effect exerced by three strictly anaerobic strains against various strains of *Clostridium perfringens* in gnotobiotic rodent intestines. Can J Microbiol 33 : 226-231

Aspects of bacterial adhesion in normal and diseased gastrointestinal tract

M Cerf

In order to avoid physical elimination, bacteria colonizing the gastro-intestinal tract of metazoans must stick in some way to intestinal tissues or contents, and in the absence of such a tight association, are known to be excreted like inert markers. This was first evidenced in a mouse model in the early sixties when differences were described between autochthonous and allochthonous bacteria [24]. Despite a host of studies during the past decades, mechanisms underlying the phenomenon of bacterial association to epithelia are still incompletely understood [3, 32].

Though bacterial adhesion to cells, one of these mechanisms, had been known for a long time [25], its importance was really evidenced with studies demonstrating that the presence of antigen K 88, a fimbrial structure of toxigenic *Escherichia coli*, was a prerequisite to the expression of the pathogenic effect of this bacterium. Toxigenic effects were shown to follow preliminary adhesion mediated by antigen K 88 [42]. Later, adhesion was shown to be a phenomenon widespread in human and animal gastrointestinal tracts and which could potentially induce numerous interactions between bacteria, epithelial cells and lamina proprial structures (eg lymphoïd and vascular tissues) [32, 43]. Bacterial adhesion appeared not to be restricted to pathological events [74] and to be one particular step in the complex process of association between bacteria and mucosal surfaces, a process which needs contact between bacterial structures and special receptors on cell surfaces [33, 74].

Some questions raised by the concept of adhesion will be considered in this paper :
- bacterial adhesins ;
- receptors (taxins) on the host cell surface ;
- experimental models ;
- the importance of adhesion in human gastro-intestinal disease.

Adhesins and colonization factors

Association of bacteria to epithelia is of prime importance in the maintenance of an adequate microecological equilibrium despite a relatively low rate of bacterial renewal and a constant shedding of bacteria due to intestinal peristalsis, cellular desquamation and mucus secretion. Adhesion is one of the factors maintaining this association.

The mechanisms for adherence may be purely physical [14] but stabilization occurs only if a strong binding is established between bacterial adhesins and mucosal receptors (eg glycocalyx). Fimbriae (pili) but also other bacterial surface structures may be involved in this process [15].

Fimbriae

Most studies have been devoted to pathogenic *Escherichia coli* in which different types of fimbriae have been described (Table 1). Fimbrial structures are best shown by electron microscopy after ruthenium red staining.

Table 1. Main fimbrial adhesion factors (from [14])

Adhesion factor	Target
K 88 (F4)	Piglet
K 99 (F5)	Calf, sheep, piglet
CFA I (F2)	Man
CFA II (F3) (CS1, CS2, CS3)	Man
PCF 8775 (CS4, CS5, CS6)	Man
CFA III	Man (?)
AFA 2230	Man
PCF E. coli 0159 : H4	Man

For some adhesion factors a heterogenous structure has been evidenced [14, 17, 44, 66]. CFA II has been demonstrated to be made of 3 immunologically different surface components called CS1 (colonizing surface antigen 1), CS2, CS3. CS3 is considered to be always present on strains expressing CFA II and could be the main adhesion factor of this complex. Some studies indicate that adhesion may not only be related to fimbrial subunits, but also to a minor product of the operon of the gene. The absence of CS1 and CS2 does not seem to be a hindrance to adhesion, and the actual role of these 2 adhesins is unsettled.

A heterogenous structure has been evidenced for another antigen (PCF 8775) with 3 components called C4, C5, C6. Whether other adhesins also have a complex composition is unknown.

Adhesion factors yield a filamentous structure made of 0.5-1 μm long fine rods with an outer diameter of 7 nm fixed onto the bacterial surface.

They are polymeric structures resulting from heliceal arrangements made of about 100 peptide subunits with a molecular weight from about 12,000 to 30,000 D. These peptide subunits include a high proportion of hydrophobic amino-acids. A high degree of homology has been described between some of these peptide chains (eg between CFA I and CS2) [14, 44].

• Fimbriae are mostly encoded by high molecular weight plasmids [29, 54, 55, 72]. Thus, genes coding for CFA I have been localized on a 58 MD plasmid. These genes code for a propeptide including a signal sequence which allows for transport towards the outer membrane of the bacterium [14].

• The plasmid coding for CFA II possesses components specific for CS1, CS2 and CS3. The regulation for the expression of these different peptides is actually unknown. The relationship between fimbrial expression and the serotype of *E. coli* is an important area for research [73].

As a whole, much remains to be learned about chemical composition and synthesis of fimbriae, which can be encoded not only by plasmids but also by chromosomal DNA.

Non fimbrial structures may also be involved in adhesion processes.

• Lipoteichoic acids have been implicated [14].

• Fibrillar non fimbrial structures have been described in the process of surface adhesion of entero-pathogenic *Escherichia coli* [45].

• Recently, so called fimbriosomes have also been demonstrated in hyperadhesive mutant strains. Fimbriosomes are 10 nm rounded structures closely associated with fimbriae and are made of a 28 kD protein which may contain the determinants of adherence of type I fimbriae [2].

• Whether some of these structures potentiate the effect of fimbriae or act independently is still unclear.

Mechanisms for adhesion

Physical factors have been implicated in adhesion, insofar as the hydrophobic structure of fimbriae could enhance bacterial adhesion to the cell membranes [14].

The importance of chemical factors was demonstrated by experiments with low concentrations of mannose which showed carbohydrate mediated adhesion inhibition for type I fimbriae. Although mannose has been the gold standard for studies in this field, other carbohydrates may be similarly involved in the process of adhesion : L-fucose, D-galactose, sialic acid, N-acetyl-D-glucosamine and polyosides or lectins have to be considered [27, 69]. This suggests that adhesion may take place through binding of adhesins to carbohydrate components of the cell surface and specially the cell glycocalyx. Thus, one has to consider the cell receptors.

Mucosal receptors

Attachment of *Escherichia coli* to urothelial cells has been demonstrated to be mediated by diverse glycolipids containing GAL-α 1 — 4 GAL-β moieties. Whether such receptors are also present on intestinal cells is unclear. A recent work [82] indicated that attachment to isolated HT29 or isolated human colonic cells may be due to this type of receptor, together with mannose sensitive

receptors. GAL-α 1 — 4 GAL β receptors appeared to be irregularly distributed among intestinal cells. Whether this adherence was confined to brush borders or also involved basolateral membranes is unclear. Obviously, if selective, attachment to brush border receptors would be of great pathophysiological relevance.

Mannose sensitive receptors are ill defined. Studies in a mouse model indicate that these receptors may be composed of one or more glycoproteins.
However the data are confusing because extracellular material could also be
involved in the apparent aggregation of bacteria to cells. This material could
be part of the mucus components rather than a cellular structure per se.
Moreover competition has been shown between some mucus glycoproteins
and brush border components [62, 81, 82].

A receptor on human erythrocytes has been identified for CFA I pili [62].
Upon extraction from erythrocyte membranes, this receptor was shown to be
a protein with apparent molecular weight of 26 KD which was bound by wheat
germ agglutinin, indicative of a high sialic acid content.

Confusing data may also be due to methodological bias, as some bacterial
strains may either express mannose-resistant AF/R1 pili or mannose sensitive
type I pili, depending on culture media used in the experiments [17]. So that
agglutination to mucus glycoproteins via one of these structures might be a
phenomenon independent of and different from proper adhesion to ileal,
colonic or Peyer's patch M cells.

Glycoproteins aggregating piliated RDEC-1 may originate from goblet
cells, and as a matter of fact, large amounts of mannose are found in the
link peptide region of mucus molecules [23, 31, 71].

Secretory IgA contained in or mixed with mucus glycoproteins does not
seem to be involved in the aggregation of E. coli by mucus, and cannot be
considered as a potential receptor for bacterial adhesins. The actual effect of
IgA may be through blocking or aggregating fimbriae by a non-carbohydrate
dependent mechanism.

Abnormalities in the composition of mucus may also be involved by inter-
fering in the process of bacterial association to epithelial cells (see below).

Specificity and distribution of receptors are likely to account for the selec-
tivity of bacterial tropisms which guide microorganisms towards various ecological niches along the digestive tract. Thus fibronectin which serves as a
receptor for Streptococcus pyogenes on buccal epithelial cells hinders adhesion of E. coli to the same cell type [34].

Cellular receptors are genetically determined as evidenced in piglets inoculated with K 88 fimbriated E. coli. Piglets lacking the specific receptors are
highly resistant to E. coli infection, this resistance being related to an autoso-
mal recessive gene [70]. Similar observations have been made in mice [41]
and on membranes of human urothelium where type and quantity of specific
glycolipid receptors are genetically determined.

Developmental changes also play a role in the bacterial receptivity of host
cells. Buccal cells of newborn infants are poor binders of streptococci, whereas the reverse is already found in 3-day old children [17]. Similarly, resistance to K 99 antigen develops with host age in the calf [65] and chicken [1].

New or abnormal receptors are expressed on cell surfaces following viral infection. Thus, cells previously resistant to bacterial colonization may become susceptible during or after viral infection [67].

In conclusion, variations of surface receptors whether genetic or acquired, trapping of bacteria in mucus secretions, and blocking of bacterial adhesins by secretory antibodies are factors which allow growth and regulation of the bacterial flora. However they also participate in the host's defense against bacteria, so that paradoxically, they must also be considered as parts of a complex system which can lead to physical and immunological eradication of bacteria [1, 30, 51].

Experimental models

Experimental models for studies on bacterial adhesion are badly needed.

For a long time, adhesion (agglutination) of *E. coli* to erythrocytes has served as a most useful instrument allowing detection of fimbrial structures. Adhesion of *E. coli* to different types of cells (eg human, bovine or guinea pig erythrocytes), allows differenciation between mannose sensitive and/or mannose-resistant adhesion [26, 28].

These characteristics are summarized on Table 2.

Table 2. Hemagglutination profile of diverse fimbrial adhesins (from [17])

RBC	Adhesion factor			
	PILI 1	CFA I	CFA II	PCF 8775
Human	Mannose (+)	Mannose (−)	0	Mannose (−)
Bovine	0	Mannose (−)	Mannose (−)	Mannose (−)
Chicken	Mannose (+)	Mannose (−)	Mannose (−)	?
Guinea-Pig	Mannose (+)	0	0	0

Adhesion to HeLa or HEP cells, both cultivated cell lines, may not be representative of in vivo conditions [8].

HT29 cells have been used for *E. coli* in binding studies using mannose or GAL-α 1-4 GAL-β as inhibitors [83]. Binding was inhibited both by α-methylmannoside or globo-tetraosyl-ceramide. A strong mannose-sensitive adhesion was demonstrated due to type I fimbriae. Fimbriae specific for GAL-α 1-4 GAL-β receptors caused only a loose adhesion to extracellular material.

Buccal epithelial cells allow easy and non invasive sampling [10]. Cells may be obtained by gentle scraping of the mucosa and iterative sampling may be performed in the same subject for longitudinal studies. Host and bacterial specificities are easy to assess through « cross over » studies. Unfixed or formalin fixed cells have equally been used for quantitative studies by optical or electron microscopy. Adhesion was evaluated by the percentage of buccal cells bearing adherent bacteria. Although buccal cells have been shown to fix various bacteria including enterobacteria, studies on intestinal microflora demand other « physiological » models.

Recent studies have focused on bacterial adhesion to isolated human intestinal and human fetal cells. Organotypic cultures have also been assayed. Colonic cells were used in a study on transformed non adherent bacteria in which either the *pil* region coding for type 1 fimbriae or the *pap* region encoding for assembly of P fimbriae had been inserted [83]. Colonic cells were obtained from surgical specimens and isolated using EDTA (1,5 mM). However it was not clear whether adhesion was to cell surface or baso-lateral membranes and the pathophysiological significance of the phenomenon was not unequivocally assessed.

Small intestinal cells from duodenal biopsies or organotypic cultures provide a most attractive model [49, 50]. In these studies EPEC strains were shown to adhere both to brush borders and baso-lateral membranes. In the presence of 0.5 % D-mannose, adhesion was specifically restricted to brush borders. When bacteria were cultivated with duodenal biopsies maintained in organotypic culture media they formed microcolonies which stuck to a high percentage of the brush border surfaces. This was associated with distorsion of the microvilli which showed aspects of effacement or cup-like projections or elongation. When followed up for 3-12 h, attachment appeared to be progressive. Bacteria first adhered to intact brush borders ; then microvillus lesions and vesiculations appeared and finally there was total villus effacement with the characteristic features found in EPEC induced diarrhea.

In these models, adhesion depends on multiple factors such as local pH, temperature, duration of incubation, and size of bacterial inoculum [17]. For CFA I or CFA II, adhesion is enhanced when pH is 6.8 ; optimal temperature may vary between 20°C ans 37°C ; optimal incubation time may vary between 20 min and 12 h ; bacterial inoculum has to be $< 10^8$/ml. Adhesion index is highest with duodenal cells and decreases with cells taken from more distal intestinal segments.

Bacterial adhesion varies among individuals [17]. Ten sets of human enterocytes were taken from 1 to 24 month old children. In children aged 1 to 3 months, no adhesion was seen. In children 1 to 2 years old, adhesion indices varied, depending on the bacterial strains tested and on the types of enterocytes tested. Similar results were obtained with adult specimens [49, 50].

This suggests that quantitative evaluations require studies with different adherent strains and different sets of target cells.

Deep freezing in dimethyl-sulfoxide and fetal serum did not significantly impair adhesion, suggesting that this procedure does not alter surface receptors.

Variations can be found in the same set of enterocytes. No more than 55 % of cells yield bacterial adhesion, and less than 20 % yield a significant adherence (> 4 bacteria/cell). These variations indicate that the distribution of receptors is inhomogeneous among cells.

Bacteria pathogenic for animals do not always adhere to human enterocytes [19] and human pathogenic *E. coli* bearing CFA I do not adhere to rabbit enterocytes. In contrast bacteria bearing CFA II (CS1 + CS3) showed significant adhesion to the same cell type in both species.

In brief, despite the sophisticated models now available, there are still great methodological difficulties in studying bacterial adhesion in vitro. Furthermore, data obtained in vitro should not be directly extrapolated to in vivo situations [1, 51]. As discussed earlier, precise knowledge of adhesion in vivo is obscured by competition between specific cell receptors and chemical components present on mucus molecules, on various lectins and even on common substances such as alimentary proteins [1, 51]. Experimentally, adhesion can be mimicked by mucus fixed on multiwell polystyren tissue culture plates.

In order to obviate these difficulties, in vivo experimental models have been designed [13]. Experiments were performed in gnotobiotic animals using 2 different strains of *E. coli*, 1 bearing and 1 lacking K 88 antigen. These experiments allowed to differentiate between loose association to diverse intestinal structures and true adhesion to epithelial cells. Such models are not fitted for clinical situations where events present with a much greater complexity than in gnotobiotic animals.

Adhesion in gastrointestinal disease

Despite these drawbacks, the pathological significance of bacterial adherence has been widely studied in acute diarrheal disease in animals as well as in humans [55, 57].

Many acute diarrheal diseases have been related to the presence of adherent *E. coli* [14, 66].

Non fimbrial adhesion has also been implicated in acute diarrheal disease due to EPEC. In this type of bacterial diarrhea, adhesion is not mediated by fimbriae but may be due to the direct sticking of the bacterial envelope to the cellular membrane.

More recently, non fimbrial fibrillar structures have been described and could have a role in the process of adhesion [50]. Microvillous effacement observed with EPEC strains [16, 80] alters cellular brush border function, but bacterial invasion does not take place [76]. Only some specific bacterial serotypes appear to be involved, due to their unique ability to express adhesion factors [60].

Bacterial adhesion, clearly a prime factor in the development of acute diarrheal disease, has been shown to play a role in some types of protracted diarrhea. This has been evidenced in acute tropical sprue and in protracted infantile diarrhea, as shown by cultures or electron microscopic examinations of duodenal biopsies [12, 78]. Simultaneous secretion of toxins was not demonstrated. Aspects of bacterial adhesion with microvillous effacement were also found in protracted diarrhea in infants [64].

Whether bacterial adhesion might play a pathophysiologic role in tropical sprue is not established [46, 47]. Because of the benefit of antibiotics in this disease and the presence of specific toxigenic strains, abnormal adhesion of bacteria to the small intestine is an attractive working hypothesis but has still not been clearly assessed. Epithelium associated bacterial overgrowths, as described in the tropics [58], have not been unequivocally related to tropical sprue.

Adhesion in chronic disease and normal subjects is certainly an important topic though little is known of the relationships between the normal commensal flora and the intestinal mucosa. A significant quantity of bacteria can be grown from small intestinal biopsies or small intestinal surgical samples [5, 58, 61]. Compared to the diverse intraluminal flora, this epithelium associated flora was generally limited to only 1 or 2 bacterial species. Bacterial concentrations were generally lower by 1 or 2 log 10 units when compared with intraluminal populations (counts rarely exceeded 10^3-10^4 cfu*/g tissue). In a group of patients with protein-calory malnutrition, bacterial counts were significantly higher and reverted to normal after malnutrition had been controlled [58].

Whether this epithelium associated flora is really adherent or only in loose contact with cellular or extracellular structures of the intestinal wall is a matter of discussion but has been rarely assessed in man [32]. In hypochlorhydric patients in whom intraluminal colonization has been known for a long time [22], cultures coupled with electron microscopic examination of duodenal or jejunal biopsies yielded growths in the order of 10^{5-6} cfu/g fresh tissue. But when comparing both methods, true adhesion could be demonstrated in only less than half of the patients whose tissue cultures had shown significant bacterial growths [11, 48] (Fig. 1).

When duodenal or jejunal biopsies were taken from different points, the bacterial population appeared to be equally distributed along a given segment of the gastro-intestinal tract [43]. More studies are needed to assess the potential significance of adherent bacteria in the upper gastro-intestinal tract.

Epithelium associated bacteria are found in the duodenum and jejunum of patients with immunodeficiency syndromes or with α-chain disease [35, 84] where malabsorption may be improved by antibiotics. Though indirect, such observations may be indicative of a pathological role of the epithelium associated flora. In the absence of a normal immunological barrier, bacterial proliferation may be enhanced, and bacteria may have direct access to the brush border glycocalyx and have damaging effects. In such conditions surface receptors to lipid components of the bacterial cell wall likely play an important role [53].

Since 1983, adherence of *Campylobacter pylori* (CP) to the antral mucosa has been of great interest to gastroenterologists. The possible pathophysiological role of CP in duodenal ulcers and type B gastritis has been an exciting subject [21]. Adhesion of CP to gastric mucosa has been shown to be selectively located at intercellular junctions although the structure mediating adhesion is actually unknown [37]. Whether CP adhering to the mucosa is really the pathogenic agent or only an innocent by-stander is still unclear, as bacterial associations with gastric epithelia have been evidenced in many other spe-

---

* Colony-forming-unit.

cies : association of spirochetes to the fundic mucosa of the dog, association of lactobacilli to the rat rumen, association of multiple species (including anaerobes) to the gastric pouch of ruminants. None of these associations have been clearly shown to be pathological.

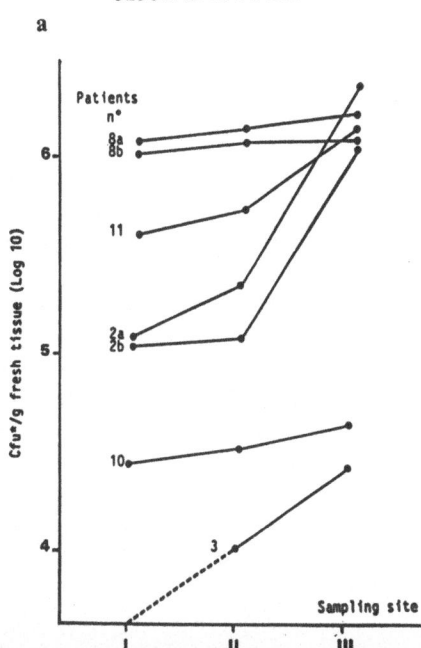

Fig. 1. a Bacterial counts from serial jeju-nal biopsies in patients with Billroth II gastrectomy. Biopsies were taken at 10 cm intervals and homogenized. Aliquots of homogenates were diluted from 10^{-1} to 10^{-6} and cultivated on various media. Results are expressed as log10 counts per g fresh tissue. b *E.coli* associated with jejunal microvilli in a patient with Billroth II gastrectomy

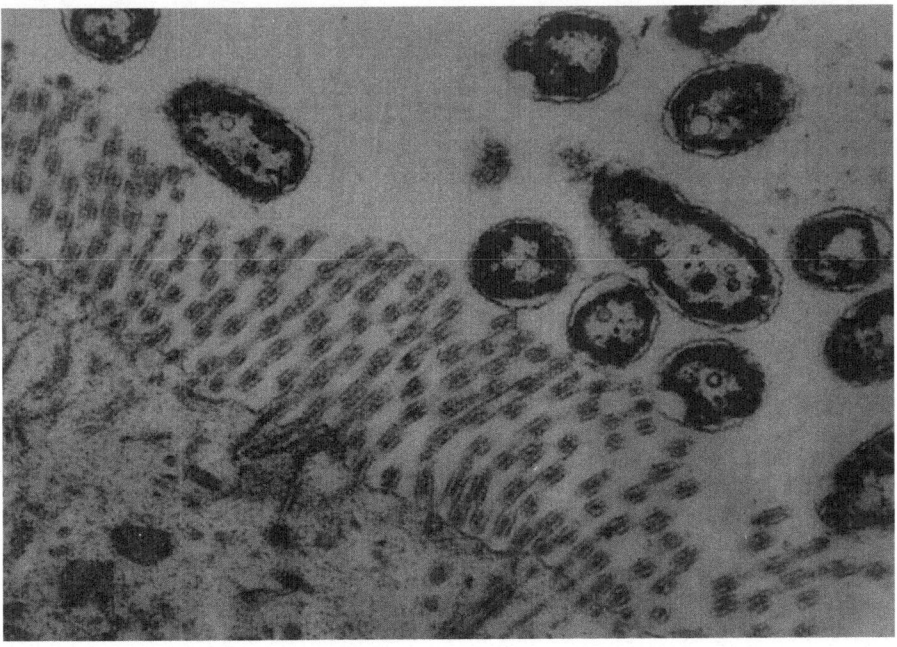

The tremendous bacterial concentration in the lower gastro-intestinal tract has raised many questions about its potential role in chronic bowel disease, especially in inflammatory bowel disease. Nevertheless little work has so far been performed in man because of methodological difficulties. Technical (and ethical) limitations are numerous and do not allow a clearcut evaluation of the importance of bacterial adherence to the colonic or rectal mucosa in inflammatory bowel disease.

Adhesion of spirochetes to rectal epithelial cells has been demonstrated through electron microscopy in animals and man [59, 75], although the mechanisms of adhesion are unknown. Rectal spirochetes have been shown in 5-15 % of symptomless subjects and in a somewhat higher proportion of homosexuals. Other bacterial associations with lower gastro-intestinal mucosae have been described in different animal species [18, 68].

Despite the high predominance of anaerobic bacteria in the lower gastro-intestinal tract, their relationship with colonic or rectal epithelia has not really been studied. True adhesion of fastidious anaerobes may be impossible due to inadequate oxygen pressure or redox potential in the cellular microenvironment, although fastidious anaerobes may be trapped in mucus or fecal material. Only adhesion capacities of facultative anaerobes (eg enterobacteria) have been studied, and only few studies have been devoted to this subject. Adherence of *E. coli* and other enterobacteria has been demonstrated by scanning electron microscopy and by quantitative cultures of biopsy specimens [6, 36] (Fig. 2). Much like within the upper gastro-intestinal tract, bacteria were evenly distributed along different colonic segments, forming micro-

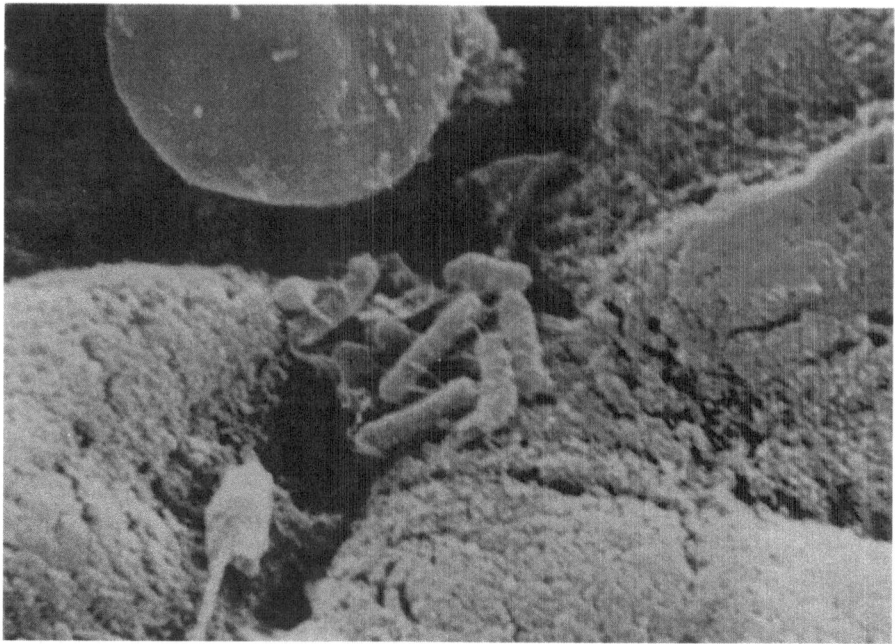

Fig. 2. Coliforms adhering to colonic mucosa in a normal subject

colonies sticking to the mucosa. Bacterial density varied from 10^6 to 10^7 cfu/g fresh tissue ; the species found were mostly *E. coli* or *Klebsiella* and sometimes *Bacteroïdes* and *Clostridia*.

When performed in patients with Crohn's disease, tissue cultures yielded no qualitative differences but showed a trend towards higher bacterial counts (10^8 cfu/g fresh tissue) compared with normals [20]. The relevance of such observations to the pathophysiology and natural history of inflammatory bowel disease needs clarification.

Other studies [9] have looked at adherence of *E. coli* in ulcerative colitis. Two cell models using either HeLa or human buccal cells have shown an abnormally high incidence of adhesive mannose-resistant *E. coli* in patients with ulcerative colitis. Differences between ulcerative colitis patients and normal controls were highly significant. But whether these phenomena are directly involved in the pathogenesis of the disease or a consequence of previous mucosal impairment is unsettled. Abnormal adhesion in inflammatory bowel disease may be enhanced by mucosal alterations, abnormal glycoprotein material [63] or unmasking of epithelial receptors secondary to cellular abnormalities. Class II MHC expression on epithelial cell surfaces is found in various diseases (eg in colonic Crohn's disease) and may be of great importance [40]. Enterocytes expressing class II MHC can process and present antigens to immunocompetent cells [38]. Thus, specific or abnormal immunological reactions may favour abnormal bacterial adhesion or abnormal reactions of the mucosa to the presence of bacteria, leading to the pathological aspects of inflammatory bowel disease.

Bacterial translocation is known to occur in diverse situations, and preliminary adhesion to cell structures may be a prerequisite to translocation [4]. Indirect proof is given by a high incidence of *Streptococcus bovis* endocarditis in patients with colonic adenomas. Thus, a common commensal bacterium may have pathological effects selectively in patients presenting with epithelial dysplasia, suggesting some particular form of association between *Streptococcus bovis* and abnormal cells [52]. Other translocation phenomena as seen with *Klebsiella* in immunosuppressed patients [77] or with *E. coli* in cirrhosis may also first require adhesion to intestinal cells.

In conclusion, during the last decades much work has been done in the field of bacterial adhesion. It has been focused mostly on adhesion of enterobacteria, and specially *E. coli*, yielding new insights into the mechanisms of acute diarrheal disease. What is now needed is investigation on the relevance of bacterial adhesion to the normal process of development and equilibrium of intestinal microflora and the role of adhesion in chronic intestinal disease. More data in this field are eagerly awaited by gastro-enterologists.

References

1. Abraham SN, Beachey EH (1985) Host defences against adhesion of bacteria to mucosal surfaces. In : Gallin JI, Fauci AS (Eds) Advances in host defence mechanisms. Raven Press, New York, pp 63-88

2. Abraham SN, Goguen JD, Beachey EH (1988) Hyperadhesive mutant of type-1 fimbriated *Escherichia coli* associated with formation of FIMH organelles (fimbriosomes). Infect Immun 56 : 1023-1029

3. Allweiss B, Dustal J, Carey KE, Edwards TF, Freter R (1977) The role of chemotaxis in the ecology of bacterial pathogens of mucosal surfaces. Nature 266 : 488-450

4. Berg RD, Garlington AW (1979) Translocation of certain indigenous bacteria from the gastro-intestinal tract to the mesenteric lymph nodes and other organs in a gnotobiotic mouse model. Infect Immun 23 : 403-411

5. Bergogne-Berezin E, Zechovski N, Cerf M, Pappo ME, Debray Ch (1973) Étude critique de 100 tubages jéjunaux. Essai de confrontation des données cliniques et bactériologiques. Pathol Biol 2 : 505-515

6. Bergogne-Berezin E, Cerf M, Gaudin B, Cazier A, Bizet J, Feldman G (1986) Recherche des phénomènes d'adhésion bactérienne dans le tube digestif chez l'Homme. Rev Inst Pasteur (Lyon) 19 : 93-103

7. Bhat P, Albert MJ, Rajan D, Ponnian G, Nathan VI, Baker SJ (1980) Bacterial flora of the jejunum : a comparison of luminal aspirate and mucosal biopsy. J Med Microbiol 13 : 247-255

8. Burke DA, Axon ATR (1987) HeLa cell and buccal cell adhesion assays for detecting intestinal *Escherichia coli* with adhesive properties in ulcerative colitis. J Clin Pathol 40 : 1402-1404

9. Burke DA, Axon ATR (1987) Ulcerative colitis and *Escherichia coli* with adhesion properties. J Clin Pathol 40 : 782-786

10. Candy DCA, Leung TSM, Phillips AP, Harries JT, Marshall WL (1981) Models for studying the adhesion of enterobacteria to the mucosa of the human intestinal tract. In : Adhesion and microorganism pathogenicity. Ciba Foundation Symposium 80, Pitman Medical, Tunbridge Wells, pp 72-93

11. Cerf M, Gaudin B, Cazier A, Barge J, Bizet J; Bergogne-Berezin E (1986) Bacterial adhesion in human upper gastrointestinal tract. Diagn Microbiol Infect Dis 5 : 285-291

12. Challacombe DN, Richardson JM, Rowe B, Anderson CM (1984) Bacterial microflora of the upper-gastro-intestinal tract in infants with protracted diarrhea. Arch Dis Child 49 : 270-277

13. Chappuis JP, Duval-Iflan Y, Ducluzeau R, Raibaud P (1985) Resistance of gnotobiotic large white and chinese piglets to in vivo attachment of a K 88 ab enterotoxigenic *Escherichia coli* strain. Reprod Nutr Dev 25 : 45-60

14. Contrepois M (1988) Les colibacilles pathogènes : adhérence et facteurs de colonisation des colibacilles entérotoxinogènes. In : Rambaud JC, Modigliani R (Eds) L'intestin grêle. Excerpta Medica, Elsevier, pp 138-150

15. Costerton JW, Irvin RT, Cheng KJ (1981) The bacterial glycocalyx in nature and disease. Ann Rev Microbiol 35 : 299-324

16. Cravioto A, Gross RJ, Scotland SM, Rowe B (1979) An adhesion factor found in strains of *Escherichia coli* belonging to the traditional infantile enteropathogenic serotypes. Curr Microbiol 3 : 95-99

17. Darfeuille-Michaud A (1987) Contribution à la recherche et à l'identification des facteurs d'adhésion d'*Escherichia coli* responsables de diarrhées ; description de nouveaux facteurs d'adhésion. Thèse de Doctorat : Sciences Université Clermont II

18. Davis CP, Savage DC (1974) Habitat, succession, attachment and morphology of segmented filamentous microbes indigenous to the murine gastro-intestinal tract. Infect Immun 10 : 948-956

19. Deneke CF, Mc Gowan K, Larson AD, Gorbach SL (1984) Attachment of human and pig (K 88) enterotoxigenic *Escherichia coli* strains to either human or porcine small intestinal cells. Infect Immun 45 : 522-524

20. Dickinson RJ, Varian SA, Axon ATR, Cooke EM (1980) Increased incidence of

coliforms with in vitro adhesion and invasive properties in patients with ulcerative colitis. Gut 21 : 787-792

21. Dooley CP, Cohen W (1988) The clinical significance of *Campylobacter pylori*. Ann Intern Med 108 : 70-79

22. Drasar BS, Shiner DDM, Mc Leod GM (1969) Studies on the intestinal flora I. the bacterial flora of the gastro-intestinal tract in healthy and achlorhydric persons. Gastroenterology 56 : 71-76

23. Drumm B, Roberton AM, Sherman PM (1988) Inhibition of attachment of *Escherichia coli* RDEC-1 to intestinal microvillus membranes by rabbit ileal mucus and mucin in vitro. Infect Immun 56 : 2437-2442

24. Dubos R, Schaedler RW, Costello RL (1963) Composition, alteration and effect of the intestinal flora. Fed Proc 22 : 1322-1326

25. Duguid JP, Gillies RR (1957) Fimbriae and adhesive properties in dysentery bacilli. J Path Bacteriol 74 : 397-411

26. Duguid JP (1964) Functionnal anatomy of *Escherichia coli* with special reference to enteropathogenic *Escherichia coli*. Rev Latinoam Microbiol 13-14 : 1-16

27. Etzler ME (1979) Lectins as probes in studies of intestinal glycoproteins and glycolipids. J Clin Nutr 32 : 133-138

28. Evans DG, Evans DJ, TJOA WS (1977) Hemagglutination of human group A erythrocytes by enterotoxigenic *Escherichia coli* isolated from adults with diarrhea : correlation with colonization factor. Infect Immun 18 : 330-337

29. Evans DG, Silver RP, Evans DJ, Chase DJ, Gorbach SL (1975) Plasmid controlled colonization factor associated with virulence in *Escherichia coli* enterotoxigenic for human. Infect Immun 12 : 656-657

30. Evans DG, De La Cabada FJ, Evans DJ Jr (1982) Correlation between intestinal immune response to colonization factor antigen I and acquired resistance to enterotoxigenic *Escherichia coli* diarrhea in an adult rabbit model. Eur J Clin Microbiol 1 : 178-185

31. Forstner GG, Forstner JF (1986) Structure and function of gastrointestinal mucus. In : Desnuelle P, Sjostrom H, Noren O (Eds) Molecular and cellular basis of digestion. Elsevier, New York, pp 125-143

32. Freter R, O'Brien PCM, Halstead FA (1978) Adhesion and chemotaxis as determinants of bacterial association with mucosal surfaces. Adv Exp Med 107 : 429-437

33. Freter R (1981) Mechanisms of association of bacteria with mucosal surfaces in Adhesion and microorganism pathogenicity. Ciba Symposium 80, Pitman Medical (Ed), pp 36-55

34. Gibbons RJ, Van Houte J (1975) Bacterial adherence in oral microbial ecology. Ann Rev Microbiol 19 : 19-44

35. Harzic M, Girard-Pipau F, Halphen M (1985) Étude bactériologique, parasitologique et virologique de la flore digestive dans la maladie des chaînes α. Gastroenterol Clin Biol 9 : 472-479

36. Hartley CL, Neumann CS, Richmond MH (1979) Adhesion of commensal bacteria to the large intestine wall in human. Infect Immun 23 : 128-132

37. Hazell SL, Lee A, Brady L, Hennessy W (1986) *Campylobacter pylori* and gastritis. Association with intercellular spaces and adaptation to an environment of mucus as important factors in colonization of the gastric epithelium. J Infect Dis 153 : 658-663

38. Heddle RJ, Shearman DJC (1979) Serum antibodies to *Escherichia coli* in subjects with ulcerative colitis. Clin Exp Immun 38 : 22-30

39. Hirschberger M, Thalee MM, Mirelman D (1977) Mechanisms of attachment by a pathogenic strain of *Escherichia coli* 0111/B4 to intestinal mucosa in pre and post weanling rats. Pediatr Res 11 : 500-510

40. Hodgson MJF, Jewell DP (1987) Immunology of inflammatory bowel disease. Clin Gastroenterol 1 : 531-545

41. Itoh K, Matsui T, Tsuji K, Mitsuoka T, Veda K (1988) Genetic control in the susceptibility of germ free inbred mice to infection by *Escherichia coli* 0115, a, c : K (B). Infect Immun 56 : 930-935

42. Jones GW, Rutter JM (1972) Role of K 88 antigen in the pathogenesis of neonatal diarrhea caused by *Escherichia coli* in piglets. Infect Immun 6 : 918-927

43. Jones GW (1977) The attachment of bacteria to the surfaces of animal cells. In : Reissing (Ed) Microbial interactions. Chapman and Hall, London, pp 133-176

44, Klemm P, Gaastra W, Mc Connel M, Smith H (1985) The CS2 fimbrial antigen from *Escherichia coli* : purification, characterization and partial covalent structure. Fems Microbiol Lett 26 : 207-210

45. Knutton S, Lloyd DR, Mc Neish AS (1987) Identification of a new fimbrial struture in enterotoxigenic *Escherichia coli* (ETEC) serotype 0148 : HL28 which adheres to human intestinal mucosa : a potentially new human ETEC colonization factor. Infect Immun 55 : 86-92

46. Klipstein FA, Short HB, Engert RF, Jean L, Weaver GA (1976) Contamination of the small intestine by enterotoxigenic coliform bacteria among the rural population of Haïti. Gastroenterology 70 : 1035-1041

47. Klipstein FA, Schenk EA (1975) Enterotoxigenic intestinal bacteria in tropical sprue. Effect of the bacteria and their enterotoxins on intestinal structure. Gastroenterology 68 : 642-655

48. Knutson NG, Mc Kee J, Welsh JP, Griffith WJ, Flournoy DJ (1982) Endoscopic cultures of the proximal gastro-intestinal tract. Gastrointest Endosc 28 : 12-18

49. Knutton S, Lloyd DR, Candy CA, Mc Neish AS (1985) Adhesion of enterotoxigenic *Escherichia coli* to human small intestinal enterocytes. Infect Immun 48 : 824-831

50. Knutton S, Lloyd DR, Mac Neish AS (1987) Adhesion of enteropathogenic *Escherichia coli* to human intestinal mucosa. Infect Immun 55 : 69-77

51. Lake AR, Bloch KJ, Neutra MR, Walker WA (1979) Intestinal goblet cell mucus release II. In vivo stimulation by antigen in the immunized rat. J Immunol 122 : 834-837

52. Leport J, Leport C, Vilde JL, Cerf M (1987) Endocardites a *Streptococcus bovis* et pathologie colique : à propos de 42 observations. Gastroenterol Clin Biol 11 : 25A

53. Linder H, Engberg I, Mattsby-Baltzer I, Jann K, Svanborg-Eden C (1988) Induction of inflammation by *Escherichia coli* on the mucosal level : requirement for adherence and endotoxin. Infect Immun 56 : 1309-1313

54. Lintermans P, Powl P, Deboeck F (1988) Isolation and nucleotide sequence of the F17-A gene encoding the structural protein of the F17 fimbriae in bovine enterotoxigenic *Escherichia coli*. Infect Immun 56 : 1475-1488

55. Mac Sweegan E, Walker R (1986) Identification and characterization of two *Campylobacter jejuni* adhesins for cellular and mucus substrates. Inf Immun 53 : 1431-1435

56. Mac Connel MM, Thomas LV, Willshaw GA, Smith HR, Rowe B (1988) Genetic control and properties of coli surface antigens of colonization factor antigen IV (PCF 8775) of enterotoxigenic *Escherichia coli*. Infect Immun 56 : 1974-1980

57. Mac Neish AS, Tuerner P, Fleming J, Evans N (1975) Mucosal adherence of human enteropathogenic *Escherichia coli*. Lancet II : 946-948

58. Nelson DP, Mata LJ (1970) Bacterial flora associated with human gastrointestinal mucosa. Gastroenterology 58 : 58-67

59. Nielsen RH, Orholm M, Pedersen JO (1983) Colorectal spirochetosis : clinical significance of the infestation. Gastroenterology 85 : 62-67

60. Peeters JE, Geeroms R, Orskov F (1988) Biotype, serotype and pathogenicity of

attaching and effacing enteropathogenic *Escherichia coli* strains isolated from diarrheic commercial rabbits. Infect Immun 56 : 1442-1448

61. Plaut AG, Gorbach SL, Weinstein L, Spanknebel G, Levitan R (1967) Studies of intestinal microflora III. The microflora of human small intestinal mucosa and fluids. Gastroenterology 53 : 868-872

62. Pieroni P, Worobec EA, Paranchych W, Armstrong GD (1988) Identification of a human erythrocyte receptor for colonization factor antigen I pili expressed by H10407 enterotoxigenic *Escherichia coli*. Infect Immun 56 : 1334-1340

63. Podolsky DK, Isselbacher KJ (1984) Glycoprotein composition of colonic mucosa : specific alteration in ulcerative colitis. J Clin Invest 77 : 1263-1271

64. Rothbaum R, Mac Adams AJ, Giannella R, Partin JL (1982) A clinicopathologic study of enterocyte adherent *Escherichia coli* : a cause of protracted diarrhea in infants. Gastroenterology 83 : 441-454

65. Runnels PL, Moon HW, Schneider RW (1980) Development of resistance with host age to adhesion of K 99 + *Escherichia coli* to isolated intestinal epithelial cells. Infect Immn 28 : 298-300

66. Sansonetti P (1987) Déterminants de virulence chez les bacilles à gram négatif. Médecine/Sciences 3 : 68-74

67. Sanford BA, Shellokov A, Ramsay MA (1978) Bacterial adherence to virus infected cells : a cell culture model of bacterial superinfection. J Infect Dis 137 : 176-181

68. Savage DC, Blumershine RVH (1974) Surface-surface associations in microbial communities populating epithelial habitats in the murine gastro-intestinal ecosystem : scanning electron microscopy. Infect Immun 10 : 240-250

69. Sharon N, Eshdat Y, Silverblatt FJ, Ofek I (1981) Bacterial adherence to cell surface sugars. In : Adhesion and microorganism pathogenicity. Ciba Foundation Symposium 80, Pitman Medical, Tunbridge Wells, pp 119-141

70. Sellwood R, Gibbons RJ, Jones GW, Rutter JM (1975) Adhesion of enteropathogenic *Escherichia coli* to pig intestinal brusch borders. The existence of two pig phenotypes. J Med Microbiol 8 : 405-411

71. Sherman PM, Boedeker EC (1987) Pilus mediated interactions of the *Escherichia coli* strain RDEC-1 with mucosal glycoprotein in the small intestine of rabbits. Gastroenterology 93 : 734-743

72. Smith HR, Willshaw GA, Rowe B (1982) Mapping of a plasmid coding for colonization factor antigen I and heat stable enterotoxin production isolated from an enterotoxigenic strain of *Escherichia coli*. J Bacteriol 149 : 264-275

73. Smith H, Scotland S, Rowe B (1983) Plasmid that code for production of colonization factor antigen II and enterotoxin production in a strain of *Escherichia coli*. Infect Immun 40 : 1236-1239

74. Sugarman B, Donta ST (1979) Specificity of attachment of certain *Enterobacteriaceae* to mammalian cells. J Gen Microbiol 115 : 509-512

75. Takeuchi A, Zeller JA (1972) Ultrastructural identification of spirochetes and flagellated microbes at the brush border of the large intestinal epithelium of the rhesus monkey. Infect Immun 6 : 1008-1018

76. Takeuchi A, Inman LR, O'Hanley PD, Cantey JR, Lushbaugh W (1978) Scanning and transmission electron microscopy study of *Escherichia coli* 015 (REDC-1) enteric infection in rabbits. Infect Immun 19 : 686-688

77. Tancrede CH, Andremont HO (1985) Bacterial translocation and gram negative bacteremia in patients with hematological malignancies. J Infect Dis 152 : 99-103

78. Tomkins AM, Drasar BS, James WPT (1975) Bacterial colonization of jejunal mucosa in acute tropical sprue. Lancet I : 59-62

79. Tzipori S, Robins-Browne RM, Gonig G, Hayes J, Withers M, Mac Cartney E (1985) Enteropathogenic Escherichia coli enteritis. Evaluation of the gnotobiotic piglet as a model of human infection. Gut 26 : 50-578

80. Ulshen MH, Rollo JL (1980) Pathogenesis of *Escherichia coli* in man. Another mechanism. N Engl J Med 302 : 99-101

81. Wadolkowski EA, Laux DC, Cohen PS (1988) Colonization of the streptomycin treated mouse large intestine by a human fecal *Escherichia coli* strain : role of growth in mucus. Infect Immun 56 : 1030-1035

82. Wadolkowski EA, Laux DC, Cohen PS (1988) Colonization of the streptomycin treated mouse large intestine by a human fecal *E. coli* strain : role of adhesion to mucosal receptors. Infect Immun 56 : 1036-1043

83. Wold AE, Thorssen M, Hull S, Svanborg EC (1988) Attachment of *Escherichia coli* via mannose or GAL α-1 GAL β containing receptors to human colonic epithelial cells. Infect Immun 56 : 2531-2536

84. Webster ADB (1976) The gut and immunodeficiency disorders. Clin Gastroenterol 5 : 323-340

Ecologic means of protection against *C. difficile* infections in gnotobiotic mice

G Corthier

It is now well established that *C. difficile* is involved in pseudomembranous colitis and in antibiotic-associated diarrhea (Bartlett et al, Larson et al). The pathogenicity of the organism is due to the production of 2 toxins : toxin A (an enterotoxin) and toxin B (a cytotoxin). Each purified toxin is able to kill animals when administred by the oral route (Lyerly et al, Sullivan et al). On the other hand, children very often harbor *C. difficile* strains in their digestive tract without any pathologic effect (Larson et al). Most of the strains are nontoxinogenic, but some of them are able to produce toxins in vitro. However, they do not affect the children's health. Infants are fully susceptible to both toxins ; indeed Libby et al noticed that when high quantities of toxins A and B were present in feces, the children developed diarrhea. In adults, toxinogenic and nontoxinogenic *C. difficile* are not found so often. Pseudomembranous colitis or antibiotic-associated diarrhea are observed when *C. difficile* reaches high levels. These observations raised the following question : are there ecological conditions which may influence the pathogenicity of toxinogenic *C. difficile* strains ?

To answer this question, an animal model was needed. In the past, the only one was the hamster treated with clindamycin (Bartlett et al, 1977). This antibiotic destroys most of the gram-positive and anaerobic microbial flora and prevents the antagonistic effect of such flora on *C. difficile* strains naturally harbored by the hamster. Then the strains multiply and produce toxins, a caecitis is observed and the animals die. This model is very close to human pathology but does not permit control of the *C. difficile* infecting strain. Furthermore, microbial interactions, even when reduced, may interfere. More recently, several models using gnotobiotic animals (hare, rat and mouse) have been developed (Dabard et al, Czuprinski et al, Wilson et al). It has been shown that axenic mice infected with *C. difficile* developed pseudomembranous caecitis. Mortality depended on the *C. difficile* strain used and on in vivo toxin production. In 1979, research was initiated in our laboratory on the pathology due to *C. difficile*. Dabard et al using the gnotobiotic hare demonstrated the pathogenicity of the bacteria. Then, using tests previously described for quantitation of the 2 toxins, we utilized gnotobiotic mice in order to study ecological conditions able to prevent toxin production and consequently protect from disease. We would now like to review the advances in the field of *C. difficile* pathogenicity and protection against the disease, using gnotobiotic mice as an experimental model.

Variations in pathogenicity according to in vivo toxin production

The aim of these investigations was to determine if *C. difficile* strains producing both toxins in vitro were pathogenic.

Various *C. difficile* strains had been isolated from healthy children, adults
with pseudomembranous colitis or antibiotic-associated diarrhea. These strains
were tested for their abilities to produce toxins in vitro. The numbers of viable vegetative organisms after 4 days of culture in an anaerobic chamber, were
similar for all strains tested. Three groups of strains were found to produce
toxins. The first group contained highly toxinogenic strains (VPI and Mara).
The second group was an intermediate one containing moderately toxinogenic
strains (strains 786, 88, Si and 660). The third group included the low toxinogenic strains (R4 and M2). There was a correlation between the amounts of
toxin A and toxin B produced by these various strains, i.e. the more toxin A
was produced, the more toxin B was detected (Fig. 1).

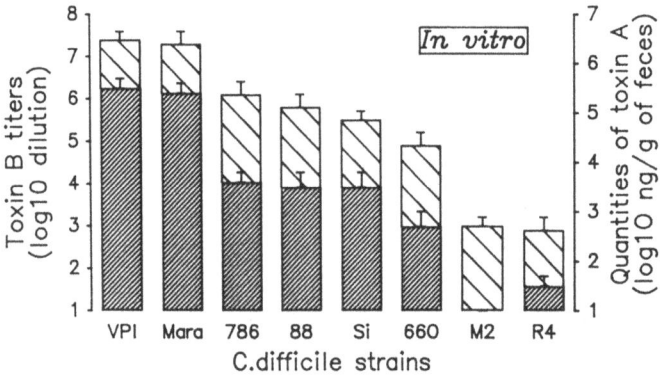

Fig. 1. Comparison of in vitro toxin A and B production by different strains of *C.*
difficile. Lightly hatched area : toxin A ; densely hatched area : toxin B

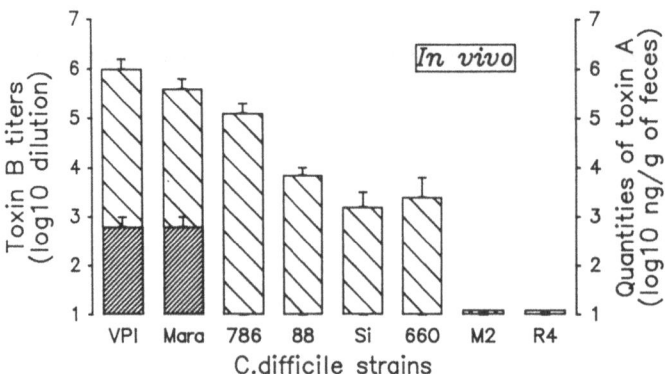

Fig. 2. Comparison of in vivo toxin A and B production by different strains of *C.*
difficile. Lightly hatched area : toxin A ; densely hatched area : toxin B

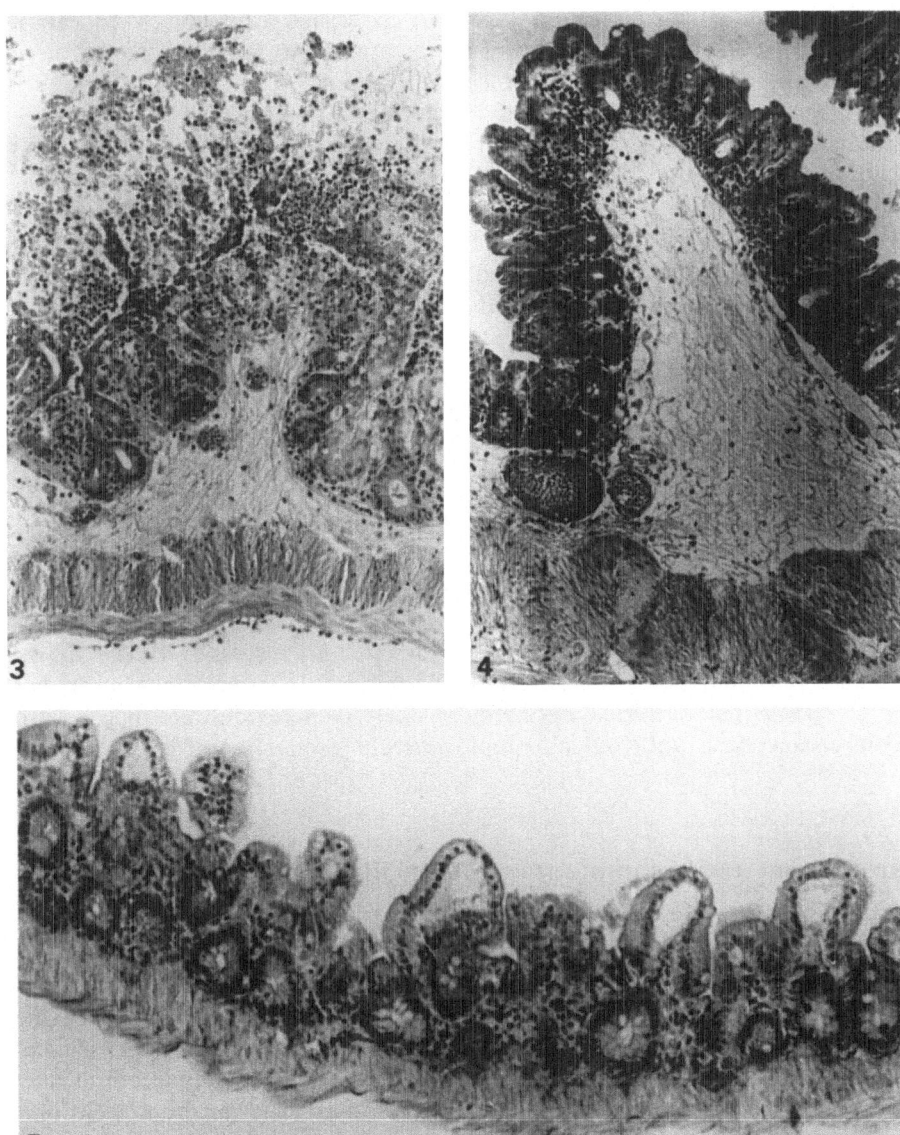

Fig. 3. Caecum of a moribund mouse 2 days after monoassociation with strain Mara (magnification 140)

Fig. 4. Caecum of a mouse 2 days after monoassociation with strain 786 (magnification 140)

Fig. 5. Caecum of an axenic mice (magnification 140)

Axenic mice were monoassociated with the different strains. Microbiologic analyses showed that, whatever the type of strain, *C. difficile* established rapidly and in large number in the large intestine. However, only strains belonging to the highly toxinogenic group were able to induce mortality. Mice harboring these strains developed a severe diarrhea and 100 % died within 2 days. No mortality was observed with the other strains.

Toxins A and B were quantitated in the caeca of the animals (Fig. 2). Moribund mice were found to have high levels of both toxins. No toxin A was detected in mice monoassociated with strains of the intermediate group, while toxin B levels still remained high. Neither toxin A nor toxin B were detected when axenic mice were monoassociated with low toxinogenic strains.

Comparison between in vitro and in vivo toxin production revealed that production of toxin A was more reduced than production of toxin B. These data suggested that production on toxin A is not inexorably linked to that of toxin B in vivo.

Histopathologic examination of the caecum was performed on mice monoassociated with each of the *C. difficile* strains. The caeca of moribund mice were characterized by pronounced epithelial abrasion and severe submucosal inflammation (Fig. 3). Mice harboring strains from the intermediate group had no epithelial caecal abrasion and a reduced inflammatory process (Fig. 4). Mice monoassociated with low toxinogenic strains had caeca similar to control mice (Fig. 5). Three weeks after challenge, the caeca of surviving mice returned to normal.

The important role of toxin A in the pathologic process is clearly shown by this study and points out the need of toxin A determinations in studies on *C. difficile* pathogenicity. Furthermore, these data demonstrate that in vitro toxin production is not a valuable tool to predict production of toxins in vivo.

Role of gut flora in protection against disease

Antagonistic effect of gut flora against C. difficile

In human and animal species, the gut flora normally exerts an antagonistic effect against multiplication of *C. difficile*. An antibiotic treatment induces destruction of various species of bacteria, and the ecologic equilibrium may be modified such that the new bacterial equilibrium may not be able to prevent *C. difficile* development. It would be of interest to know what bacteria are responsible of the antagonistic effect against *C. difficile*. Then such bacteria could be used to reduce the risk of pseudomembranous colitis. Unfortunately, such bacteria have not yet been identified although several different scientific groups have tried to identify them. Some of these bacteria probably belong to the extremely oxygen-sensitive bacteria which are difficult to cultivate even in an anaerobic chamber or have some unknown growth requirements.

Modulation of toxin production by gut flora

Axenic mice can be inoculated with a single bacterial species. Our experimental conditions using plastic isolators avoid further microbial contamination. These monoassociated mice were employed to determine if the bacteria considered had any influence on protection against pseudomembranous colitis.

As mentioned previously, children harbor *C. difficile* very often without any intestinal disorders. However, some of the *C. difficile* strains isolated from children were toxinogenic. Thus we chose to determine if the bacteria belonging to the dominant flora of the infant might prevent toxin production (Fig. 6). Highly toxinogenic *C. difficile* strains were used for the challenge performed 4 days after bacterial inoculation. All the strains tested became established at high level in axenic mice. No strain showed an antagonistic effect against growth of *C. difficile* ; however, some strains protected mice against *C. difficile*-induced death. *Escherichia coli* strains had a good protective effect whereas *Streptococcus faecalis* had a moderate effect and *Bacteroides* had none. The *Bifidobacterium* group is more complex : even within the same species, differences in protection were observed between strains.

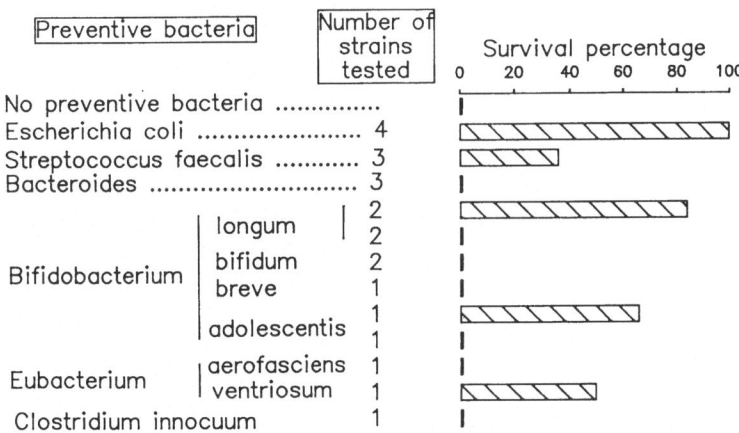

Fig. 6. Preventive effect of bacteria isolated from human dominant flora on lethality due to *C. difficile* in gnotobiotic mice

Some bacteria isolated from dominant adult human flora were also tested. Previous inoculation of mice with *Eubacterium ventriosum* induced a partial protection of the mice against pseudomembranous caecitis. However mice inoculated with *Clostridium innocuum*, *Eubacterium aerofasciens* or *Bacteroides vulgatus* and then challenged with *C. difficile* died within 3 days postinfection. Thus one cannot a priori say that a particular bacterial genus or species would prevent mortality. Strains must be tested individually. There is a clear need for understanding the mechanism of protection in order to develop an in vitro test for screening.

In dead or dying mice large amounts of toxin A and B could be detected in their caeca, whereas in surviving mice no toxin A and low levels of toxin B were detected. This effect persisted for more than 1 month. Protection thus seemed to be related to the decrease of toxin production.

Role of the yeast *(Saccharomyces boulardii)* in protection against disease

Using the above described gnotobiotic mouse model, we looked for other ecologic means of protection against experimental pseudomembranous colitis, including use of the yeast *S. boulardii*.

The protective effect

Saccharomyces boulardii is a mesophilic, nonpathogenic yeast used in many countries presumptively to prevent diarrhea and other gastrointestinal diseases associated with antibiotic use. Recent work, using hamsters treated with clindamycin (Toothaker and Elmer) revealed that *S. boulardii* could decrease mortality in this pseudomembranous colitis model. The aim of this work was to assess the ability of *S. boulardii* to prevent mortality and toxin production in the gnotobiotic mouse. In some experiments, axenic mice received a single injection by the orogastric route of a suspension containing 10^{10} viable cells/ml. For continuous treatment, the *S. boulardii* suspension was given with drinking water and replaced daily. *C. difficile* challenge, made 4 days after the beginning of each *S. boulardii* treatment, was performed with the lethal VPI strain. Given with drinking water, *S. boulardii* protected mice from *C. difficile* mortality (Fig. 7). The number of yeasts in the fecal contents was found to be 100 times lower after a single oral ingestion than after continuous treatment and the protection conferred was poor (Fig. 7). Killing the yeast by different

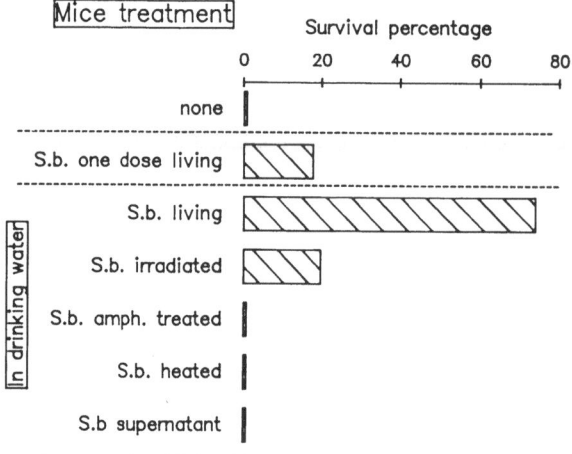

amph. : amphotéricine B

Fig. 7. Effect of *S. boulardii* (S.b.) treatment on mortality due to *C. difficile* in gnotobiotic mice

procedures : heat inactivation, gamma irradiation or amphotericin B treatment, led to the loss of protective effect suggesting that the yeast must be living during transit in the digestive tract (Fig. 7).

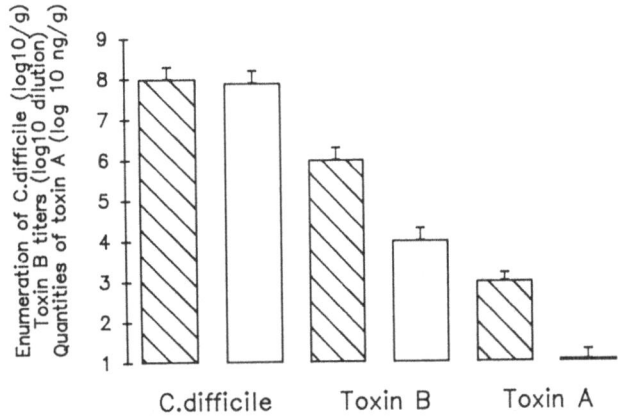

Fig. 8. *C. difficile* enumeration, cytotoxic titer and amounts of toxin A in caeca of mice treated by *S. boulardii*. Hatched area : dead or dying mice ; empty area : living mice

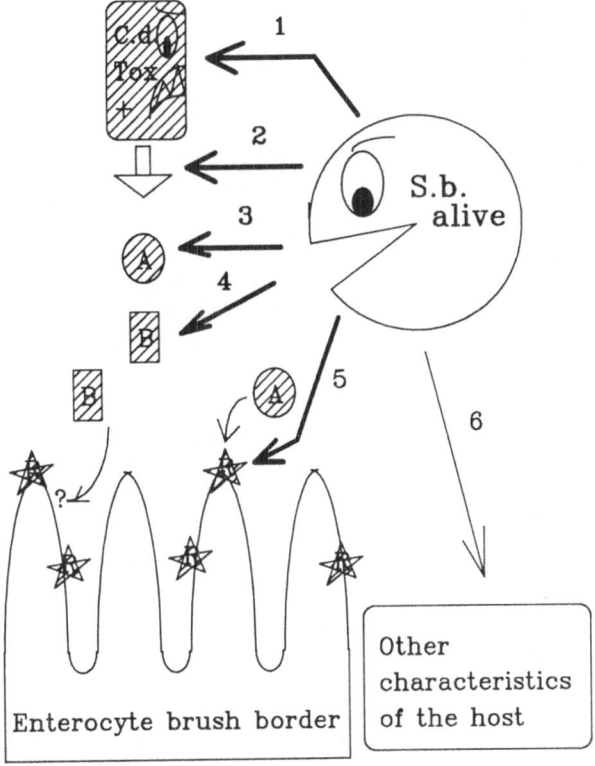

Fig. 9. Scheme of hypotheses on the mechanism of action of *S. boulardii* (S.b.)

Whatever the *S. boulardii* treatment, toxin A and B quantities were always high in dying mice. In contrast, quantities were highly reduced in surviving animals, suggesting that *S. boulardii* prevented toxin production. The effect was more pronounced on toxin A than on toxin B (Fig. 8).

Search for the mechanism of action

Different trials were undertaken to elucidate the mechanism of the protective effect of *S. boulardii*. The various hypotheses are illustrated in Fig. 9.

Hypotheses 1 : S. boulardii antagonizes growth of *C. difficile* in vivo. *S. boulardii* was not found to prevent in vitro or in vivo multiplication of *C. difficile* ;
Hypotheses 2 : S. boulardii prevents toxin production in vivo. *S. boulardii* did not block toxin production in vitro ; however it could prevent production in vivo by an indirect effect (hypothesis 6) ;
Hypotheses 3 and 4 : S. boulardii destroys toxin A or toxin B. No direct activity against toxins could be demonstrated in vitro or in vivo ;
Hypotheses 5 : S. boulardii affects toxin A binding to its receptor. Krivian et al described a sugar receptor for toxin A. *S. boulardii* contains many enzymes that metabolize saccharides, thus it could block toxin A fixation to an enterocyte. This hypothesis is now under investigation ;
Hypotheses 6 : S. boulardii may have an indirect effect by increasing the host's ability to inactivate the toxins. This hypothesis is also under investigation.

Role of food in protection against disease

This ecologic approach to protection was used in order to determine whether dietary regimens affect toxin production in vivo. Gnotobiotic mice in which the role of diet in vivo is limited to an interaction between the host and a single strain present in their intestinal tract, was an important model.

In a first experiment we compared the effect of a commercial diet to that of a semisynthetic diet (casein 22 %, cellulose 11 %, corn oil 5 %, corn starch 44 % and saccharose 11 %). Animals were adapted during 2 weeks then challenged with a *C. difficile* strain. Mice fed with the commercial diet died within 2 days while mice fed with semisynthetic diet were protected. No antagonistic effect on *C. difficile* development could be detected. In surviving mice, toxin B production was highly decreased and toxin A production was below detectable levels. This modulation of toxin production in protected animal was similar to previous observations in mice treated with *S. boulardii* (Fig. 8).

The diets were too different to determine which kind of product permit in vivo toxin production. We fed mice with different diets, directly modified from the commercial one, in which one main source of protein was used (Table 1). Low protein diet afforded good protection, but when fish meal was used as the main protein source most of the mice died after *C. difficile* challenge. As described previously, protection or death were always related to toxin production (Table 1). One can argue that the diets differed by parameters other than chemical composition (texture, taste, technologic process...). Thus we fed mice with the low protein diet and gave a fish meal solution in drinking water. The beneficial effect of the low protein diet was lost and all animals died after *C. difficile* challenge (Table 1). These data suggest that protein composition of the diet may affect pathology due to *C. difficile*. On the basis of this low protein diet supplemented by drinking water containing various sources or constituents of proteins, we tried to determine which kind(s) of product(s) allows *C. difficile* pathology i.e. : allows in vivo toxin production.

Table 1. Effect of diet on mouse survival, *C. difficile* enumeration, toxin A and B production in mice monoassociated with *C. difficile*. In *parentheses* standard error of the mean

Type of diet	Survival	*C. difficile*	Toxin A	Toxin B
Commercial diet	0	8.0* (0.3)	3.4* (0.3)	6.2* (0.2)
Origin of protein				
Fish-meal	14	7.9* (0.2)	3.3* (0.3)	5.8* (0.3)
Soya bean	43	8.1** (0.3)	<1**	5.8** (0.3)
Casein	65	7.8** (0.2)	<1**	4.3** (0.2)
Low protein diet	73	8.1** (0.3)	<1	4.4** (0.3)
Low protein diet + fish-meal in drinking water	0	8.0* (0.3)	3.4** (0.3)	6.2* (0.2)

* Dead mice, ** living mice
Toxin A is expressed in log10 dilution ; toxin B is expressed in log10 ng/g

Long-term protection after interruption of the protective treatment

Thus, some diets can protect mice against *C. difficile* pathology. One week after the challenge, the protective diet was changed to a nonprotective one. The mice remained healthy and modulation of toxin production still persisted. The same phenomenon was noticed using mice treated with *S. boulardii*. After removal of *S. boulardii* from drinking water, the yeast count decreased to a nonprotective level but the mice remained healthy and toxin production was still decreased. These data suggest that the protective agent (diet or yeast) was needed a short time (one week or less). The aim of this study was to determine whether non toxinogenic clones may be derived from a toxinogenic one in gnotobiotic mice protected against *C. difficile* infection and whether the nontoxinogenic clones can protect mice against the disease.

Method for rapid detection of toxinogenic and nontoxinogenic clones of C. difficile

Fecal samples containing *C. difficile* were enumerated in Petri dishes in an anaerobic chamber. Then a cellulose filter previously treated with anti-toxin A serum (pig species) was layered on surface colonies. The filter containing toxin A was then treated by another antitoxin A serum (rabbit species) and an anti-rabbit serum coupled to peroxidase. Staining with enzyme substrate revealed spots corresponding to toxinogenic clones. The non-spot-forming colonies corresponded to nontoxinogenic clones.

Emergence of nontoxinogenic clones

Gnotobiotic mice were protected by a *S. boulardii* treatment or an adequate diet. Then, one week after *C. difficile* challenge, the yeast was omitted from the drinking water or the protective diet was replaced by a nonprotective one. Search for nontoxinogenic clones of *C. difficile* was performed on fecal samples from time to time. Figure 10 shows the data obtained using the protective diet. Similar results were found with *S. boulardii* treatment. Nontoxinogenic clones of *C. difficile* could be found 2 weeks after challenge, and their number increased in 3 weeks to a level as high as that of toxinogenic clones of *C. difficile*. Production of toxins remained low throughout the 60-day expe-

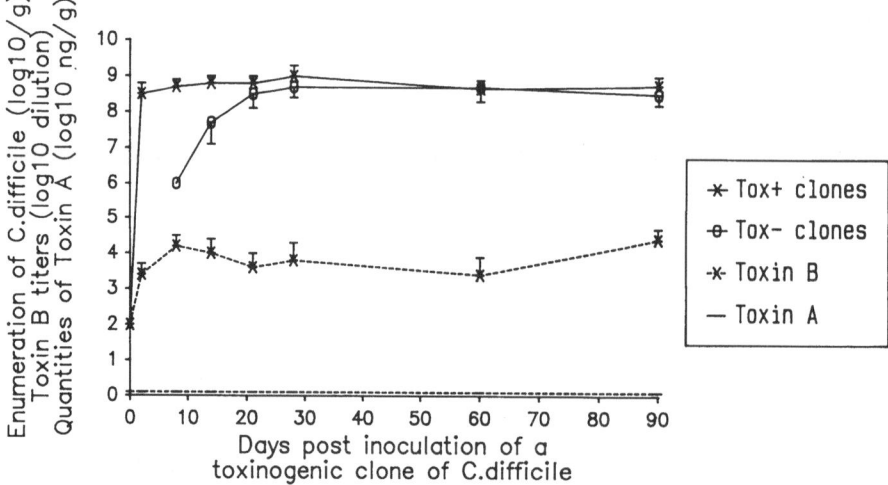

Fig. 10. Emergence of nontoxinogenic clones of *C. difficile* after treatment by a protective diet and challenge with a protective clone

riment. The nontoxinogenic clones were isolated and shown to have lost their capacities to produce toxin B and toxin A. We failed to obtain such nontoxinogenic clones in vitro from the toxinogenic clones. These data suggest that emergence of nontoxinogenic clones need some ecologic conditions that can be found only in vivo. The nontoxinogenic clones (derived from the toxinogenic clone) could be used as a preventive treatment against *C. difficile*. They exerted a highly antagonistic effect against the toxinogenic *C. difficile* (Fig. 11). The minimal time needed to obtain a protective effect was less than 18 h.

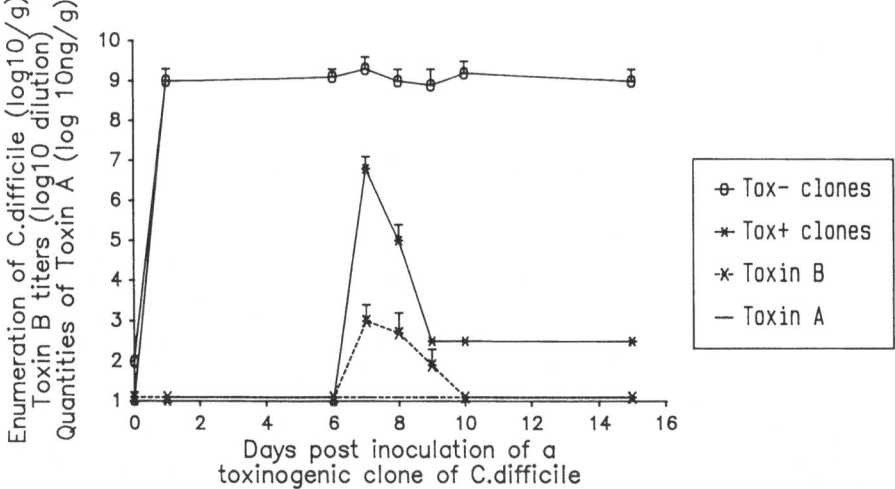

Fig. 11. Antagonistic effect of a nontoxinogenic clone of *C. difficile* on a toxinogenic clone. The toxinogenic clone was inoculated 5 days after the nontoxinogenic one

From this study it appears that *S. boulardii* or the protective diet permits mouse survival during a 10 days period following *C. difficile* challenge. Then nontoxinogenic clones emerged from the toxinogenic ones and they maintained modulation of toxin production originally due to *S. boulardii* or diet.

Conclusions

Gnotobiotic mice offer a very good animal model to study experimental pseudomembranous colitis :
. the infecting *C. difficile* strain could be chosen from a wide range of lethal toxinogenic, non-lethal toxinogenic or nontoxinogenic strains ;
. various intestinal human bacteria or yeast could be tested and used to protect mice against disease and to prevent toxin production ;
. the effects of diet composition could be tested on protection and on toxin production ;

animals could be kept without any bacterial contamination from the environment and variations in toxinogenesis of infecting strains could be followed from time to time. A better understanding of the main features of experimental pseudomembranous colitis would probably permit improved therapy or prevention of the disease.

Comment

The results presented here are derived from the following publications of the Laboratory of Microbial Ecology :
- Corthier G, Dubos F, Raibaud P (1985) Modulation of cytotoxin production of *C. difficile* in the intestinal tract of gnotobiotic mice inoculated with various human intestinal bacteria. Appl Environ Microbiol 49 : 250-252
- Corthier G, Dubos F, Ducluzeau R (1986) Prevention of *C. difficile*-induced mortality in gnotobiotic mice by *Saccharomyces boulardii*. Can J Microbiol 32 : 894-896
- Mahé S, Corthier G, Dubos F (1987) Effect of various diets on toxin production by 2 strains of *C. difficile* in gnotobiotic mice. Infect Immun 55 : 1801-1805
- Corthier G, Muller MC (1988) Emergence of nontoxinogenic clones of *C. difficile* from toxinogenic one in gnotobiotic mice. Infect Immun 56 : 1500-1504
- Vernet A, Corthier G, Dubos F, Rapine P, Parodi A (1987) Comparison of cytotoxin and enterotoxin production of various *C. difficile* strains in culture broth and in gnotobiotic mice. Proc Int Symp Gnotobiology, Versailles, France, p 446

And from publications in process :
- Vernet G, Corthier G, Ramare F, Parodi A. *Clostridium difficile* : relationship between toxin A and toxin B production and cecal lesions in gnotobiotic mice
- Elmer G, Corthier G. Modulation of *C. difficile*-induced mortality as a function of the dose and of the viability of the *Saccharomyces boulardii* used as a preventive agent.

References

Bartlett JG, Onderdonk AB, Cisneros RL, Kasper DL (1977) Clindamycin associated with colitis due to a toxin-producing species of *Clostridium* in hamsters. J Infect Dis 136 : 701-706

Bartlett JG, Chang TW, Gurwith TW, Gorbach SL, Onderdonk AB (1978) Antibiotic-associated pseudomembranous colitis due to toxin producing Clostridia. N Engl J Med 298 : 531-534

Czuprinski CJ, Johnson WJ, Balish E, Wilkins TD (1983) Pseudomembranous colitis in *C. difficile* monoassociated rats. Infect Immun 39 : 1368-1376

Dabard J, Dubos F, Martinet L, Ducluzeau R (1979) Experimental reproduction of neonatal diarrhea in young gnotobiotic hares simultaneously associated with *C. difficile* and other *Clostridium* strains. Infect Immun 24 : 7-11

Larson HE, Barclay FE, Honour P, Boriello SP (1978) Epidemiology of *C. difficile* and the aetiology of pseudomembranous colitis. Lancet I : 822-829

Larson HE, Price AB, Honour P, Wilkins TD (1981) *Clostridium difficile* and the aetiology of pseudomembranous colitis. Lancet I : 1063-1066

Libby JM, Donta ST, Wilkins TD (1983) *C. difficile* toxin A in infants. J Infect Dis 148 : 606

Lyerly DM, Lockwood DE, Richardson SH, Wilkins TD (1982) Biological activities of toxin A and B of *C. difficile*. Infect Immun 35 : 1147-1150

Lyerly DM, Saum KE, MacDonald DK, Wilkins TD (1985) Effects of *C. difficile* toxins given intragastrically to animals. Infect Immun 47 : 349-352

Sullivan NW, Pellett S, Wilkins TD (1982) Purification and characterization of toxins A and B of *C. difficile*. Infect Immun 35 : 1032-1040

Toothaker RD, Elmer GW (1984) Prevention of clindamycin-induced mortality in hamsters by *Saccharomyces boulardii*. Antimicrob Agents Chemother 26 : 552-556

Wilson KH, Sheagren JN, Freter R, Weatherbee L, Lyerly D (1986) Gnotobiotic models for study of the microbial ecology of *C. difficile* and *Escherichia coli*. J Infect Dis 153 : 547-551

Pathophysiology of *Clostridium difficile*-related intestinal disease

JC Rambaud, Ph Marteau, I Sobhani, and O Berretta

Pseudomembranous colitis (PMC) is the most severe form and probably the last stage [35] of a spectrum of acute intestinal diseases due mainly to antibiotic administration and with in situ proliferation of toxigenic *Clostridium difficile* [7, 19]. This pathogen seems also to be responsible for cases of PMC not linked to antibiotherapy [7, 42] (Table 1).

Table 1. Risk factors associated or not with antibiotic treatment in patients with *C. difficile*-related intestinal disease

Risk factors	With antibiotics	Without antibiotics
Acute/relapsing leukemia[a, g]	+	+
Aplastic anemia[a]	+	+
Bone marrow transplantation[b]	+	+
Cytotoxic chemotherapy[a, g]	+	+
Steroids[a]	+	+
Acquired immune deficiency syndrome[a]	+	
Diabetes[a, g]	+	+
Chronic renal failure[a, f, g]	+	+
Glomerulonephritis[c]		+
Critically ill burned patients[a, h]	+	
Hirschsprung's disease[d, e]		+
Inflammatory bowel disease[d]	+	+
Intestinal surgery[f, g]	+	+
Acute intestinal obstruction[f]		+
Heavy metal poisoning[f] ?		+
Ischemic cardiovascular disease[f, g] ?		+

[a] Talbot RW et al (1986) Br J Surg 73 : 457-460
[b] Yolken RH et al (1982) N Engl J Med 306 : 1009-1012
[c] Peikin SR et al (1980) Gastroenterology 79 : 948-951
[d] Brearly S et al (1987) J Pediatr Surg 22 : 257-259
[e] Thomas DFM et al (1986) J Pediatr Surg 21 : 22-25
[f] Gineston JL et al (1980) Gastroenterol Clin Biol 4 : 709-720
[g] Church JM et al (1986) Dis Colon Rectum 29 : 804-809
[h] Grube BJ et al (1987) Arch Surg 122 : 655-661

Since the recognition that this bacterium played a major role in PMC [2, 3, 23, 36], significant, although still incomplete, advance has been achieved in understanding several aspects of the mechanisms of pathogenicity of *C. difficile* in connection with alterations of the microflora ecosystem of the distal gut. This progress comes mainly from : a) studies of the role of *C. difficile* toxins in intestinal lesions and hypersecretion ; b) etiologic, epidemiologic and bacteriologic data from patients with *C. difficile*-related intestinal disease ; c) experimental studies of the effects of manipulation of the colonic micro-flora on the growth and/or toxin production of *C. difficile*. It appears that the concepts, derived from animal studies, of barrier flora and of emergence of clones of *C. difficile* with different abilities to secrete toxins are important guides to the understanding of the pathophysiology of *C. difficile*-related colitis in man.

Role of toxins in *Clostridium difficile*-related intestinal disease

Clostridium difficile is a noninvasive bacterium and exerts its noxious effects through the secretion of at least 2 toxins called A, or enterotoxin, and B, or cytotoxin. Their respective role in the pathogenesis of fluid hypersecretion and intestinal parietal lesions has been studied : a) by challenging different animal models with the two toxins ; b) by studying the effects of active immuniza-tion with toxoids A and B on these animal models ; c) by comparing in clindamycin-pretreated hamsters the virulence (i.e. lethal power and intestinal lesions) of different exogenous strains of *C. difficile* with the caecal concen-trations of toxins A and B. Some discrepancies, probably due to differences in purification procedures, doses and assays of toxins and to different animal models, hinder the synthesis of these experimental data.

In the first set of experiments, a constant finding was the accumulation of hemorrhagic fluid and the occurrence of hemorrhagic mucosal lesions in the ligated or perfused ileal loop of the rabbit challenged with toxin A, whe-reas toxin B had no or weak effects [28, 43, 45]. Similar results have been obtained with ligated ileal and colonic loops in the hamster [25, 43]. When toxins of the same source were inoculated into unligated caeca of hamsters, toxin A was lethal and induced severe ileal and caecal mucosal lesions, whe-reas toxin B was not lethal and only induced areas of focal hemorrhage [43]. A similar experiment [25] performed in hamsters with « purified » toxin A and partially purified toxin B also showed that toxin A, at concentrations found in the caecum after clindamycin treatment of this animal, caused death and produced enlarged caeca due to fluid accumulation, with moderate hemorrhage but severe mucosal damage. Toxin B challenge did not induce fluid accumu-lation ; however, at variance with the results of Taylor et al [43], it also cau-sed severe necrotic and hemorrhagic lesions of caecal mucosa although multi-focal rather than diffuse. Using ligated loops of the middle of the mouse small intestine, and another method of purification, Lönnroth and Lange [26] found that purified toxin A induced a clear and watery fluid secretion, whereas toxin B had no effect. However, crude toxin A, which was contaminated by toxin B, or a mixture of purified toxins A and B, induced a hemorrhagic secre-tion, whereas toxin A plus heat-inactivated toxin B produced a clear secretion like that induced by toxin A alone. Similarly, in further experiments, Lyerly et al [27] showed that, although purified toxin B alone had no effect when administered intragastrically to hamsters and mice, it showed a synergistic effect when simultaneoulsy administered with toxin A. Conflicting results have been reported on the lethal action and intestinal secretory effect of toxins A and B in suckling mouse assay [28, 43].

In experiments to compare the virulence and the ability of different strains of *C. difficile* to secrete in vivo toxins A and B, hamsters pretreated with clindamycin received orally nine different toxigenic strains of this bacterium and were housed individually in sterile isolators [5]. A correlation was found between virulence and toxin A, but not toxin B, levels in caecal contents ; toxins A and B were not produced in a constant ratio, by contrast with observations made in vitro. Moreover, confirming previous data, there was no correlation between virulence and production of either toxin A or B in vitro. In their experiments on the protective effects of some semisynthetic diets on the mortality of gnotobiotic mice exposed to strains of toxigenic *C. difficile*, Mahé et al [29] observed that, after the shift from the semisynthetic diets to commercial diet, the surviving animals remained healthy, although toxin B production returned to the high control levels. At the same time, toxin A levels in caecum or feces, high in controls receiving since the beginning the commercial diet, remained undetectable.

Extrapolation to humans of these findings, stressing the major role of toxin A in the pathogenicity of *C. difficile*, should be very cautious, since it is known that different animal species or even different strains of the same species can exhibit marked differences in sensitivity to the same strain of *C. difficile* [8, 12]. Unfortunately, human data on this topic are conflicting, at least in adults suffering from *C. difficile*-related intestinal disease. Some authors [46] found toxins A and B in the stools of 100 % and 95 %, respectively, of patients, while toxin A was demonstrated in only 80 % of patients harbouring toxin B in stools in another series [31]. Moreover, toxin A (like toxin B) is often found in the stools of infants with no obvious gastrointestinal distress [21, 24]. This finding could be due to the lack of epithelial surface binding sites to toxin A, whose carbohydrate sequences could be developmentally regulated [21].

Finally, the protective effects of previous immunization against toxin A or B or both on clidamycin-induced lesions in the hamster have been assessed [25]. Only hamsters immunized agains both toxins were completely protected, and nearly all of the animals protected against neither toxin or only 1 toxin died. The animals showed large caeca with fluid accumulation, which appeared less hemorrhagic in animals immunized against toxin B. The histologic pattern of the caecum was similar in the three groups, with focal necrosis of the surface epithelium and underlying inflammation and hemorrhage. These data suggest that both toxins are responsible for intestinal abnormalities and death in this experimental model. However, the authors do not exclude that immunization against toxin A was not complete and that residual toxin A, and not toxin B, was responsible for at least intestinal fluid secretion.

On the whole, toxin A appears to play a prominent role in the intestinal secretion and pathologic lesions due to *C. difficile* overgrowth, whereas toxin B may only potentiate the effects of the enterotoxin.

However, are toxins A and B entirely responsible for the enteropathogenicity of *C. difficile* ? A recent study suggests the presence in *C. difficile* cul-

ture of an enterotoxic factor distinct from toxin A and inducing nonhemor-
rhagic clear fluid accumulation in the rabbit ileal loop assay, without mucosal
damage [18]. The relationship of this new factor with the toxin A purified by
Lönnroth et Lange [26] remains to be established. Finally, heat-labile factors
that alter the myoelectric activity in rabbit isolated ileal loops were found in
the crude culture filtrate of *C. difficile* and in a high molecular weight filtration product obtained from the culture supernatant, whereas the effects of purified toxins A and B were not different from those of saline controls [20]. Noticeably, the high molecular weight factor also caused mucosal damage.

Clostridium difficile growth and toxin production

That the growth of *C. difficile* and its ability to secrete toxins are in large
part determined by its microbiologic environment in the distal part of the intestine is supported by the etiology and epidemiology of *C. difficile*-related intestinal disease in humans and by experimental data obtained in several animal
models.

Lessons from etiology and epidemiology of C. difficile-*related colitis*

Several clinical observations strongly suggest that toxigenic *C. difficile* proliferation is favored by alterations of the bacterial ecosystem of the distal gut.
      Current or recent treatment by antibiotics is by far the most frequent cause
of *C. difficile*-related colitis [7]. All antibiotics, with the exception of parenteral aminoglycosides which are not excreted into bile, may be responsible for
the disease [17, 42]. However, these last data do not take into account the
relative consumption of the various antibiotics. On this basis, β lactamase stable
penicillins are 8 times, cephalosporins 40 times, and lincosamides 70 times more
frequently associated with *Clostridium difficile*-related colitis than narrow-
spectrum penicillins [1].
      These epidemiologic data fit very well with the impact of the various anti-
biotics on the fecal flora, which in turn is related to both their in vitro activity on strictly anaerobic colonic bacteria and their fecal levels. The latter are
governed by the antibiotic absorption rate in the small intestine and by the
biliary excretion of the native drug (Table 2). Experiments in healthy volunteers have shown that alterations in fecal anaerobic microflora by 7-day administration of a sulbactam pivoxil and bacampicillin combination (decreased
counts of *Bacteroides*, elevated population level of *Clostridia*) were maximal
at day 5 [39], which is the most usual delay of onset of antibiotic-related intestinal disease (Fig. 1).
      An intriguing feature is that several antibiotics often implicated in *C.
difficile*-related colitis, such as ampicillin, are active in vitro against this bacterium [7]. In hamsters with ampicillin-induced colitis, β lactamase activity
increased in the distal gut to a level sufficient to destroy ampicillin, allowing
C. difficile overgrowth and toxin production [37]. This finding implies that

Table 2. Pharmacokinetics, in vitro activity and in vivo impact on dominant fecal flora in relationship with the risk of diarrhea of different antibiotics

Antimicrobial	In vitro activities on anaerobic colonic flora	Pharmacokinetics			Impact on anaerobic fecal flora	Risk of diarrhea
		absorption	biliary excretion	fecal levels		
Penicillins						
Ampicillin	+++	++	-	++	++	++
Azlocillin	+++	-	+	++	++	++
Bacampicillin	++	++++	+	+	+	+
Piperacillin	+++	-	+	++	++	++
Cephalosporins and other beta lactam antibiotics						
Cefoperazone	+++	-	++	+++	+++	+++
Cefoxitin	+++	-	+	++	+++	+++
Aztreonam	+	-	+	++	-	++
Imipenem	+++	-	+	+	-	+
Macrolides						
Clindamycin	+++	++	++	+++	+++	+++
Erythromycin	++	+++	++	++	+++	++
Nitromidazoles						
Metronidazole	+++	+++	+	-	-	-
Tetracyclines						
Doxycycline	++	++	++	++	+++	++
Quinolones						
Norfloxacin	+	+++	+	++	-	-
Ciprofloxacin	+	+++	+	++	-	-

+ = low, ++ = moderate, +++ = high, - = nil
Adapted from Bergan T (1986) Scand J Infect Dis (Suppl) 49 : 91-99 ; and from Nord CE et al (1986) Scand J Infect Dis (Suppl) 49 : 64-72

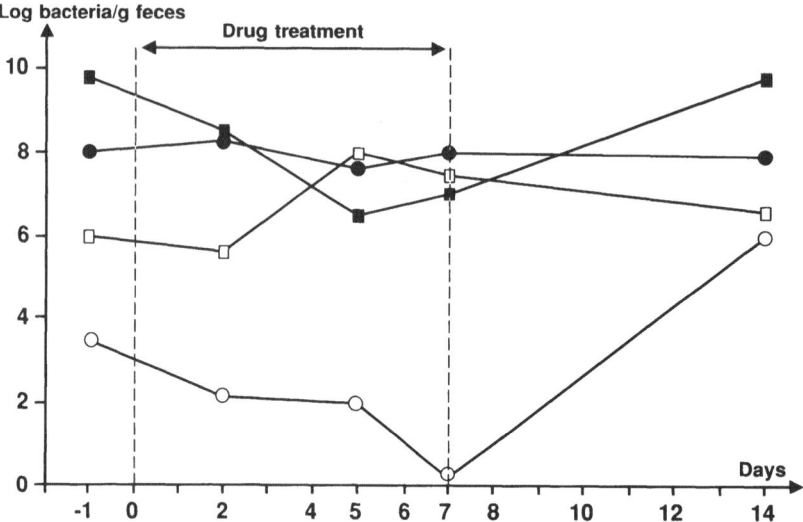

Fig. 1. Effect of sulbactam pivoxil and bacampicillin combined administration on the intestinal anaerobic microflora of 15 healthy volunteers. Median number per g feces of the following bacteria is indicated : *Bacteroides (closed squares), Clostridium (open squares), Bifidobacteria, Lactobacilli (closed circles)* and *Veillonella (open circles)*. Adapted from Jövall et al (1986) Scand J Infect Dis (Suppl) 49 : 72-84

the recovery of the barrier flora to *C. difficile* initially depressed by ampicillin was slower than the proliferation of the pathogen. A similar mechanism could explain the occurrence of *C. difficile*-related intestinal disease after cessation of antimicrobial therapy with an antibiotic to which this pathogen is susceptible. Indeed, hamsters given oral vancomycin, which is very active against *C. difficile* and many other colonic anaerobic bacteria, expire with *C. difficile*-induced caecitis at 5-7 days after antibiotic withdrawal [16]. Well-documented cases of *C. difficile*-related intestinal disease have also been observed during treatment with clindamycin, to which the bacterial strains were very susceptible. There is no current evidence of « clindamycinase » and the mechanism of this observation remains unknown.

In combination with antibiotic administration and sometimes alone, other factors can induce *C. difficile*-related colitis (Table 1). The large majority can be classified according to 2 physiopathologic mechanisms : immunodeficiency or at least debilitating conditions, and fecal stasis. In both cases, alterations of the intestinal ecosystem are very likely, although not always proven.

One or more relapses often occur after vancomycin therapy of PMC and are due to incomplete eradication of *C. difficile*, which probably persists mainly in spore form. A clinical observation pointing to the major role of the persistance of disturbancies of the colonic microflora in this inability to eradicate *C. difficile* is the definitive cure of some cases of antibiotherapy-linked PMC by enemas with normal stool homogenates after several relapses treated by repeated courses of oral vancomycin [7, 38].

The indigenous or exogenous source of the strains of *C. difficile* which proliferate in the distal intestine under the above-mentioned etiologic conditions remains disputed [32]. In hospitals, clusters or outbreaks of PMC or toxin-positive diarrhea have been repeatedly reported, especially in surgical, oncologic and pediatric wards [7]. In most [7, 15] but not all [7, 13, 33] of these, an environmental source of *C. difficile* was suspected because positive cultures were obtained from toilet-seats, bedpans, floors, hands and stools of asymptomatic personnel working in areas in which these disorders appeared. In another study, the use of unsterilized sigmoidoscopes may have contributed to the appearance of multiple cases in a single ward [7]. However, evidence of the true nosocomial origin of *C. difficile* infection relies on typing of the bacteria isolated from the patients and their environment. Unfortunately, there is a lack of simple and widely available typing systems for *C. difficile*. Methods used include serogrouping, bacteriophage and bacteriocin-typing, crossed immunoelectrophoresis, polyacrilamide gel electrophoresis (PAGE), electroblot transfer, incorporation of ^{35}S-methionin into bacterial proteins, followed by PAGE and autoradiography, plasmide profile analysis and whole-cell DNA restriction endonuclease profile analysis [13, 22, 33, 34, 40, 41]. The results of such studies are conflicting. In some outbreaks, a unique type of *C. difficile* was found [13, 34], whereas in others several different types were identified [22, 40].

Carriage in stools of *C. difficile* is observed in 0-3 % of healthy adults, and in a much greater proportion of young infants [7]. Moreover, low numbers of *C. difficile* would escape detection with present methods. This could explain the sporadic cases of *C. difficile*-related intestinal disease and, together with direct patient-to-patient cross-infection [33], non-nosocomial outbreaks in hospitals. However, as already stressed, most infants harbouring *C. difficile* or its toxins remain healthy, possibly because of the lack of intestinal toxin A receptors [21].

Experimental data

Several in vitro and in vivo studies have clearly shown that bacterial and yeast interactions and diet manipulation may promote or repress the growth and toxin secretion of *C. difficile*.

To investigate the influence of normal intestinal flora in implantation and proliferation of *C. difficile*, Borriello and Barclay [4] measured the growth and cytotoxin production of *C. difficile* in emulsions prepared from feces of various patients and healthy controls. Special attention was paid to the assessment of the relative roles of viable bacteria and cell-free components (e.g. pH, volatile fatty acids, residual antibiotics) in modulating the growth of *C. difficile*. Growth of an exogenous strain of *C. difficile* and its cytotoxin production were inhibited in all fecal emulsions from healthy adults, although at a low degree in three cases. Viable bacteria were very probably responsible for this inhibition, since it disappeared when homogenates were sterilized. Loss of the inhibitory effets of stool homogenates of patients with *C. difficile*-mediated diarrhea on growth of the « indigenous » strain of *C. difficile* was found in 79 % of cases. Similar observations were made when the same strain of *C. difficile* as that used for experiments in healthy subjects was inoculated on emulsions of the caecal content of an untreated hamster and of a clindamy-cin-treated hamster maintained individually in sterile isolators. Homogenates from the untreated animal inhibited the growth and cytotoxin secretion of *C. difficile*, whereas the same homogenate sterilized by filtration and the caecal homogenate of the clindamycin-treated animal were not inhibitory [4]. Many attempts have been made to identify the individual components of the fecal flora that are antagonistic to *C. difficile* in vitro. Unfortunately, the experimental conditions related poorly to those prevailing in vivo, and in most cases the antagonistic effect on *C. difficile* of the various bacteria studied was in fact attributable to the acidic pH produced [4]. Thus, in vivo experiments in suitable animal models are necessary to a better understanding of bacterial or yeast interaction with *C. difficile* growth and toxin production. Conventional and gnotobiotic young hares, gnotobiotic rats and mice, and antibiotic-treated hamsters were used in these studies.

Clostridium difficile is the causal agent of neonatal and often fatal diarrhea in conventional and gnotobiotic young hares. Some associated bacteria are able to enhance or repress the pathogenic effect of *C. difficile* in these animals. *C. difficile* is very often associated with *C. perfringens* and *C. tertium* in the feces of conventional diarrheic young hares [12] and these 3 bacteria were never simultaneously found in the feces of healthy animals. When a strain of these three bacterial species was monoassociated with axenic young hares, all were found at a high population level in the feces, but only *C. difficile* was pathogenic [12]. However, when a strain of *C. perfringens* or of *C. tertium*, or both, was associated with *C. difficile*, diarrhea and death occured earlier than in hares monoassociated with *C. difficile*, whereas the association of *C. perfringens* and *C. tertium* was non-pathogenic [12]. Moreover, *C. difficile* became established more rapidly when associated with *C. perfringens*. Thus, although *C. perfringens* and *C. tertium* are non-pathogenic for the gnotobiotic young hare, they enhance the pathogenicity of *C. difficile* in this animal model.

On the other hand, a microbial barrier can prevent neonatal diarrhea induced by *C. difficile* in young conventional and gnotobiotic hares [14]. The fecal flora of gnotobiotic mice inoculated with the caecal content of a healthy young hare (a transfer allowing monitoring of the absence of *C. difficile* in the inoculum) was transferred into conventional hares immediately after birth, control hares not being inoculated. Animals were maintained in a closed building with standardized room temperature and artifical lighting. No mortality was observed during the first 15 days of life in inoculated hares, whereas 35 % of the noninoculated ones died of diarrhea during this period, the difference being highly significant. Selective enumeration of *C. difficile* made at various intervals showed no or mild and transient presence of this bacterium in the feces of inoculated hares and of noninoculated animals having survived at least 18 days. By contrast, *C. difficile* was present in all dead hares studied. The same barrier flora protected until weaning 2 young gnotobiotic hares challenge-exposed 4 days later with *C. difficile* and *C. perfringens* [14]. These findings are in agreement with those previously reported by Fekety et al [15] on the protective role of caecal homogenate of conventional hamster on the *C. difficile*-related lethal caecitis induced by the withdrawal of vancomycin treatment in this animal. Attempts to isolate a combination of pure microbial strains that could exert against *C. difficile* the same antagonism in gnotobiotic hares or vancomycin-treated hamsters as that exerted by the entire barrier flora were unsuccessful, although a combination of 4 or 6 isolates from hares had a barrier effect against *C. perfringens* and, to a lesser extent, *C. difficile* in gnotobiotic mice [14]. However, recently, a diassociation of *Clostridium cocleatum* (whose implantation necessitated previous inoculation of a variant strain of *Bifidobacterium bifidum*, which was eliminated by *C. cocleatum* growth) and a gram-negative, spore-forming semi-circular rod probably belonging to the *Fusobacterium species* completely protected gnotoxenic mice from a toxigenic *C. difficile* challenge [6] by completely eliminating this bacterium.

A yeast, *Saccharomyces boulardii*, can also exert an antagonistic action on the intracaecal growth of *C. difficile* [30] and on the related lethal caecitis induced by a clindamycin challenge in the hamster [30, 44]. In the experiments of Massot et al [30], oral administration of the yeast for 13 days, starting three days prior to the antibiotic treatment, reduced the mortality rate and large intestinal lesions compared to control animals having received clindamycin only. Likewise, *C. difficile* levels in the caecum and colon were reduced, although still higher than those found in control animals which had not received the antibiotic (Fig. 2). The significant reduction of mortality and frequency of caecal lesions by *S. boulardii* was confirmed independently in the same animal model [44]. The protection achieved by the yeast was all or none. As *S. boulardii* survives in, but does not colonize the intestine, daily administration was

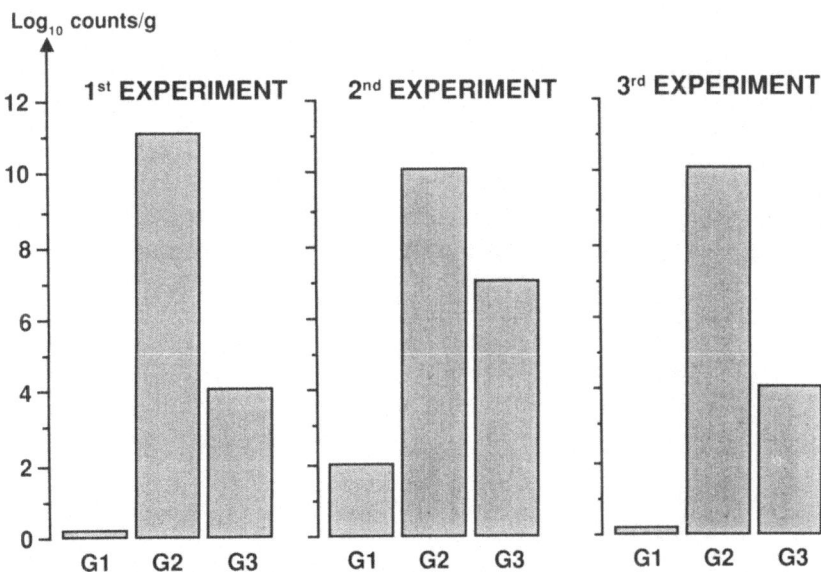

Fig. 2. Effect of *Saccharomyces boulardii* on *C. difficile* caecal population level in the clindamycin-challenged hamster model. G 1 = control group ; G 2 = clindamycin group (1 mg/kg) ; G 3 = clindamycin (1 mg/kg) + *S. boulardii* (5×10^9/day) group. Each column represents the result in a pool of 2 hamsters, and the experiment was repeated three times

necessary to obtain a sustained concentration (> $10^{6.5}$ colony forming units/g) in the caecal contents. However, all animal surviving at the end of yeast administration, i.e. 10 days after clindamycin treatment, remained healthy subsequently.

Modulation of *C. difficile* implantation and growth is not the only way to modify the pathogenic effect of this bacterium. Indeed, several experiments have shown that some agents can modify toxin secretion by *C. difficile* without noticeably altering its population level in the caecum or stools. Gnotobiotic mice, which developed acute caecitis and died 2 days after the inoculation of a cytotoxigenic *C. difficile* strain, were completely protected by the inoculation four days earlier of a strain of *Escherichia coli* or of a strain of *Bifidobacterium bifidum* (Fig. 3) isolated from the feces of healthy neonates [9]. The population level in caecal contents of *C. difficile* was not significantly different from that of the reference mice (dead mice diassociated with *C. difficile* and other bacteria than *E. coli* and *B. bifidum*) in *B. bifidum*-harbouring mice, and only slightly (30 %) although significantly lower in *E. coli* inoculated animals. In contrast, the caecal toxin B titers were 1000 times lower in mice inoculated with *E. coli* or *B. bifidum* than in reference mice, the differences being highly significant. The study of the kinetics of cytotoxin titers in stools of mice diassociated with *B. bifidium* and *C. difficile* showed little variations during more than one month [9]. Further studies showed that toxin A production was nil in the protected animals (G Corthier, personal communication).

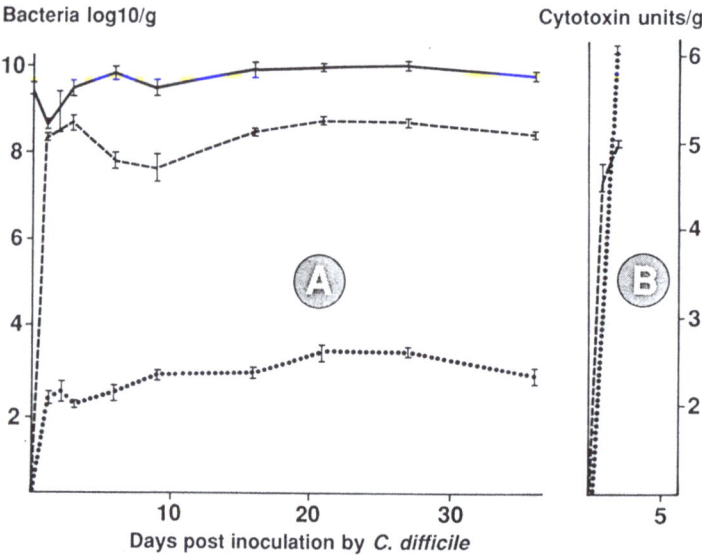

Fig. 3. Kinetics of cytotoxin production and bacterial establishment in gnotobiotic mice after inoculation of *C. difficile*. *A* : mice previously inoculated with *B. bifidum* ; *B* : mice axenic before inoculation with *C. difficile*. *Vertical bars* represent standard deviation. *Solid line* : bacterial counts of *B. bifidum* ; *broken lines* : bacterial counts of *C. difficile* ; *dotted lines* : cytotoxin titer. Adapted from G Corthier *et al* (1985) Appl Environ Microbiol 49 : 250-252

It is known that a first challenge with a nontoxigenic strain of *C. difficile* can protect cefoxitin-treated hamsters against a toxigenic strain of this bacterium [47]. The possible mechanisms of this protective effect have been studied by Corthier et al [8]. Cytotoxin production by human and hare strains of *C. difficile* were compared both in vitro in a broth culture and in vivo in intestinal contents of gnotobiotic rats and mice. The human strain produced about 1000 times more cytotoxin than the hare strain, both in vitro and in vivo, although the population levels of the two strains were similar in both experimental conditions. Ninety per cent of gnotobiotic rats and 100 % of gnotobiotic mice established with the human strain died within 3 days, whereas no mortality was observed with the hare strain. Moreover, previous establishment of the hare strain protected from challenge with the human strain. This effect was related to a marked and sustained decrease of cytotoxin fecal levels, which could not be explained only by the slight, although significant, barrier effect of the hare strain on the human one. It is noticeable that the cytotoxin levels observed in the diassociated animals were not only lower than in those monoassociated with the human strain, but also under those observed in animals monoassociated with the hare strain. Thus, cytotoxin production by both human and hare strains was modulated by their association in gnotobiotic mice.

Another study performed by the same group [10] suggests that a modulation of cytotoxin production could, beside the above-mentioned effect on *C. difficile* population level [30], explain in part the protective action of *S. boulardii* on *C. difficile*-related colitis. The continuous administration of a dense suspension of *S. boulardii*, so that its fecal count consistently reached $10^{9.4}$ yeasts/g, to gnotobiotic mice inoculated with *C. difficile* decreased the mortality rate from 100 % in control animals to 44 % in treated ones. No difference was found in the counts of *C. difficile* in the two groups, whereas a marked and highly significant lowering of intracaecal cytotoxin levels, and also of toxin A concentrations (G Corthier, personal communication) was observed between surviving animals protected by *S. boulardii* and those from both groups which died. Interestingly, the cytotoxin levels remained low after withdrawal of *S. boulardii* administration, in spite of the decrease of yeast count in feces. However, at the time of yeast treatment interruption, *C. difficile* was still pathogenic for germ-free mice placed in contact with protected mice. In these contact animals, which all died within two days, high *C. difficile* numbers and low *S. boulardii* counts were observed in stools, whose cytotoxin content was high [10].

Microorganisms are not the only factors modulating toxin secretion by *C. difficile*. Mahé et al [29] have shown that the composition of the diet influences enterotoxin and cytotoxin secretion and mortality in axenic mice monoassociated with 2 toxigenic strains of *C. difficile*. When fed with a commercial diet, 100 % of the animals inoculated with strain VIP died three days after inoculation, and both enterotoxin and cytotoxin were produced in their caeca. With 3 semisynthetic diets, the first balanced, the second protein-rich and the third carbohydrate-rich, the mortality rates 5 days after *C. difficile* inoculation were 0, 30 and 10 %, respectively. Population levels at day 2 of strain VIP in the caecum or stools were high and not statistically different, compared with animals receiving the commercial diet. However, cytotoxin and ente-

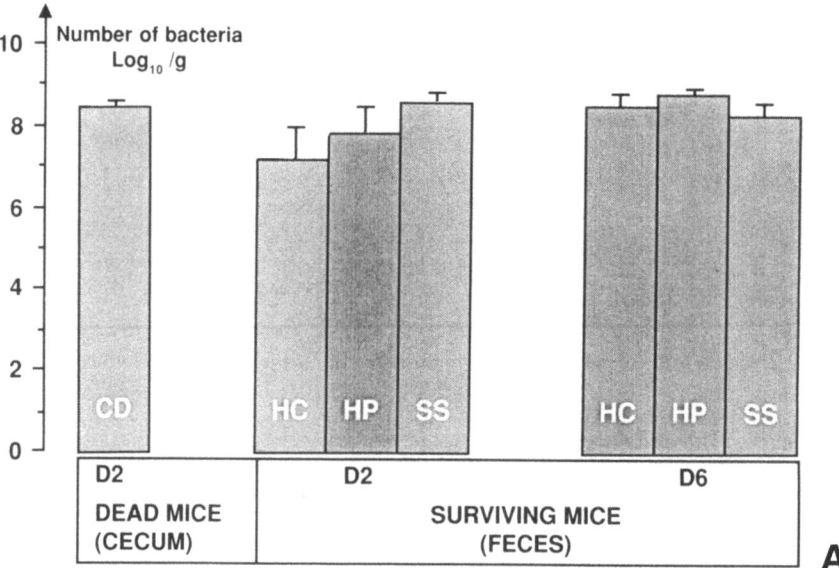

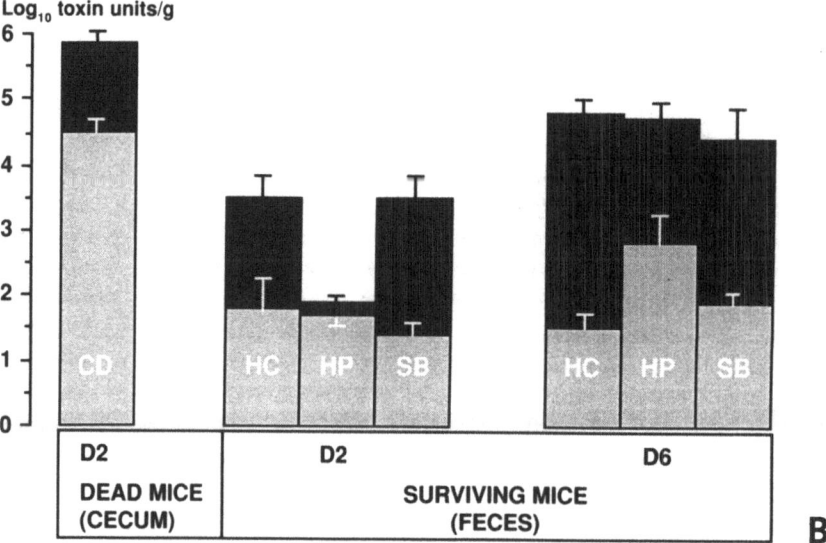

Fig. 4 A, B. Effects of diet on population levels of *C. difficile* and on toxin production in gnotobiotic mice monoassociated with *C. difficile*. CD : animal receiving commercial diet ; HC, HP and SB = groups of mice given high-carbohydrate, high-protein or balanced semisynthetic diet, respectively. *Vertical bars* = standard deviation. **A** The type of diet has no influence on the total counts of *C. difficile* at days 2 (D2) and 6 (D6) after starting the diet ; **B** cytotoxin *(black columns)* and enterotoxin *(grey columns)* production is significantly reduced at day 2 with HC, HP and SB diets compared to CD, and this decrease persists at day 6. Adapted from S Mahé et al (1986) Microecol Therapy 16 : 209-216

rotoxin levels in the 3 groups of animals receiving the semisynthetic diets were significantly lower than those observed in mice fed with the commercial diet. Similar, but not identical, results were observed with the other strain of *C. difficile*. When surviving animals implanted with either strains were shifted from their semisynthetic diet to the commercial one, none died and enterotoxin was never detected ; cytotoxin production remained low and stable with the first strain, whereas it increased gradually with the second one to levels found in the control group fed with the commercial diet. Figure 4 depicts the results of a similar experiment also performed by Mahé et al and showing the significant decrease of both cytotoxin and enterotoxin levels, without alteration of *C. difficile* population level, in mice receiving semisynthetic diets, compared to animals fed with the commercial diet.

Thus, the previous establishment of several species of bacteria, such as some strains of *E. coli, B. bifidum* and *C. difficile* itself, or the continous administration, starting before *C. difficile* implantation, of *S. boulardii*, can reduce the mortality rate of different gnotobiotic rodents associated with toxigenic strains of *C. difficile*, by reducing toxin secretion, whereas the population level of the bacteria remains unchanged or only slightly decreases. The same mechanism explains the protection of gnotobiotic mice by some semisynthetic diets. Moreover, in some of these experimental models, the protection persisted after stopping the administration of the protective agent. In gnotobiotic mice protected by *S. boulardii* or a semisynthetic diet, Corthier and Muller [11] have shown that this protection was related to the emergence of nontoxigenic clones of *C. difficile* from the toxigenic one. These clones became established at the same level as the toxigenic one after about 1 month. In those protected mice bearing nontoxinogenic clones, no enterotoxin production could be detected and cytotoxin titers were markedly reduced, although the population level of the toxigenic clone did not change. These nontoxinogenic clones and clones that produced intermediate levels of toxins in vivo did not revert to toxin production, even after repeated culture in vitro. Also, no nontoxigenic clone could be obtained from toxigenic ones in vitro (G Corthier, personal communication). The nontoxigenic clones were shown to arise from a single toxigenic clone and were indentical to that clone in metabolic patterns and antibiotic sensitivity tests. Finally, mice fed with a nonprotective diet challenged with a nontoxinogenic or intermediate clone remained healthy and no toxin production could be detected in their feces. Interestingly, as in the experiments with hare and human *C. difficile* strains mentioned above [8], these mice were protected against further challenge with toxigenic strains of *C. difficile*. However, this protection was now mainly due to a strong antagonistic effect of the nontoxinogenic clones against the toxigenic ones.

In conclusion, all etiologic, epidemiologic and experimental data concerning *C. difficile*-related intestinal disease point out disturbances of the intestinal bacterial ecosystem as the main cause of this pathologic entity. Several protective factors have been discovered, antagonizing either *C. difficile* growth or toxin production or both. However, the mechanisms of these antagonisms are still poorly understood and their elucidation could lead to new preventive and curative treatments of these frequent and sometimes life-threatening complications of antibiotherapy, immunosuppression and fecal stasis.

Acknowledgements. The authors are very grateful to D Ehrlich for typing the manuscript and I Wohnlich for her bibliographic assistance.

References

1. Aronsson B, Mollby R, Nord CE (1985) Antimicrobial agents and *C. difficile* in acute enteric disease : epidemiologic data from Sweden. J Infect Dis 151 : 476-481

2. Bartlett JG, Chang TW, Gurwith M, Gorbach SL, Onderdonk AB (1978) Antibiotic-associated pseudomembranous colitis due to toxin producing clostridia. N Engl J Med 298 : 531-534

3. Bartlett JG, Moon N, Chang TW, Taylor N, Onderdonk AB (1978) Role of *Clostridium difficile* in antibiotic-associated pseudomembranous colitis. Gastroenterology 75 : 778-782

4. Borriello S, Barclay FE (1986) An in vitro model of colonisation resistance to *Clostridium difficile* infection. J Med Microbiol 21 : 299-309

5. Borriello SP, Ketley JM, Mitchell TJ (1987) *Clostridium difficile* : a spectrum of virulence and analysis of putative virulence determinants in the hamster model of antibiotic-associated colitis. J Med Microbiol 24 : 53-64

6. Boureau H, Decré D, Popoff M, Bertocci A, Su WJ, Bourlioux P (1988) Isolation and identification of microflora resistant to colonization by *C. difficile*. XIII International Symposium on Intestinal Microecology. Porto Conte, sept 11-14. Abstract Book

7. Chang TW (1985) Antibiotic-associated injury to the gut. In : JE Berk (Ed) Bockus Gastroenterology, 4th edition, vol 4. Saunders, Philadelphia, pp 2583-2592

8. Corthier G, Dubos F, Raibaud (1986) Ability of two *Clostridium difficile* strains from man and hare to produce cytotoxin in vitro and in gnotobiotic rodent intestines. Ann Inst Pasteur/Microbiol 137 B : 113-121

9. Corthier G, Dubos F, Raibaud P (1985) Modulation of cytotoxin production by *Clostridium difficile* in the intestinal tracts of gnotobiotic mice inoculated with various human intestinal bacteria. Appl Environ Microbiol 49 : 250-252

10. Corthier G, Dubos F, Ducluzeau R (1986) Prevention of *Clostridium difficile* induced mortality in gnotobiotic mice by *Saccharomyces boulardii*. Can J Microbiol 32 : 894-896

11. Corthier G, Muller MC (1988) Emergence in gnotobiotic mice of nontoxigenic clones of *Clostridium difficile* from a toxigenic one. Infect Immun 56 (in press)

12. Dabard J, Dubos F, Martinet L, Ducluzeau R (1979) Experimental reproduction of neonatal diarrhea in young gnotobiotic hares simultaneously associated with *Clostridium difficile* and other *Clostridium* strains. Infect Immun 24 : 7-11

13. Delmee M, Bulliard G, Simon G (1986) Application of a technique for serogrouping *Clostridium difficile* in an outbreak of antibiotic-associated diarrhoea. J Infect 13 : 5-9

14. Dubos F, Martinet L, Dabard J, Ducluzeau R (1984) Immediate postnatal inoculation of a microbial barrier to prevent neonatal diarrhea induced by *Clostridium difficile* in young conventional and gnotobiotic hares. Am J Vet Res 45 : 1242-1244

15. Fekety R, Kim KH, Brown D, Batts DH, Cudmore M, Silva J Jr (1981) Epidemiology of antibiotic-associated colitis. Isolation of *Clostridium difficile* from the hospital environment. Am J Med 70 : 906-908

16. Fekety R, Kim KH, Batts BH (1980) Studies on the epidemiology of antibiotic-associated *C. difficile* colitis. Am J Clin Nutr 33 : 2527-2532

17. Finegold SM (1986) Anaerobic infections and *Clostridium difficile* colitis emerging during antibacterial therapy. Scand J Infect Dis (Suppl) 49 : 160-164

18. Gianfrilli PM, Piemonte F, Giuliano M (1988) Isolation of a new enterotoxic fac-

tor from *Clostridium difficile* (Abstr). International Congress for Infectious Diseases. Rio de Janeiro, April 17-21. Abstract Book

19. Goligher JC (1984) Pseudomenbranous colitis (PMC)-antibiotic-associated colitis (AAC). In : Surgery of the anus, rectum and colon. Baillière Tindal, Eastbourne, pp 1024-1027

20. Justus PG, Martin JL, Goldberg DA (1982) Myoelectric effects of *Clostridium difficile* : motility-altering factors distinct from its cytotoxin and enterotoxin in rabbits. Gastroenterology 83 : 836-843

21. Krivan HC, Clark GF, Smith DF, Wilkins TD (1986) Cell surface binding site for *Clostridium difficile* enterotoxin : evidence for a glycoconjugate containing the sequence Galα1-3 Glcβ1-4glnNac. Infect Immun 53 : 573-581

22. Kuijper EJ, Oudbier JH, Stuifbergen WN, Jansz A, Zanen HC (1987) Application of whole-cell DNA restriction endonuclease profiles to the epidemiology of *Clostridium difficile*-induced diarrhea. J Clin Microbiol 25 : 751-753

23. Larson HE, Price AB (1977) Pseudomembranous colitis : presence of clostridial toxin. Lancet II : 1312-1314

24. Libby JM, Donta ST, Wilkins TD (1983) *Clostridium difficile* toxin A in infants. J Infect Dis 148 : 606

25. Libby JM, Jortner BS, Wilkins TD (1982) Effects of the two toxins of *Clostridium difficile* in antibiotic-associated caecitis in hamsters. Infect Immun 36 : 822-829

26. Lonnroth I, Lange S (1983) Toxin A of *Clostridium difficile* : production, purification and effect in mouse intestine. Acta Pathol Microbiol Immunol Scand [B] 91 : 395-400

27. Lyerly DM, Saum KE, McDonald DK, Wilkins TD (1985) Effects of *Clostridium difficile* toxins given intragastrically to animals. Infect Immun 47 : 349-352

28. Lyerly DM, Lockwood DE, Richardson SH, Wilkins TD (1982) Biological activities of toxins A and B of *Clostridium difficile*. Infect Immun 35 : 1147-1150

29. Mahé S, Corthier G, Dubos F (1987) Effect of various diets on toxin production by two strains of *Clostridium difficile* in gnotobiotic mice. Infect Immun 55 : 1801-1805

30. Massot J, Sanchez O, Couchy R, Astoin J, Parodi AL (1984) Bacteriopharmacological activity of *Saccharomyces boulardii* in clindamycin-induced colitis in the hamster. Arzneimittelforschung/Drug Res 34 : 794-797

31. Mollby R, Aronsson B, Nord CE (1985) Pathogenesis and diagnosis of *Clostridium difficile* enterocolitis. Scand J Infect Dis 46 (Suppl) : 47-56

32. Mullingan ME (1984) Epidemiology of *Clostridium difficile*-induced intestinal disease. Rev Infect Dis 6 (Suppl) : 222-228

33. Pierce PF Jr, Wilson R, Silva J Jr (1982) Antibiotic-associated pseudomembranous colitis : an epidemiologic investigation of a cluster of cases. J Infect Dis 145 : 269-274

34. Poxton IR, Aronsson B, Mollby R, Nord CE, Collee JG (1984) Immunochemical fingerprinting of *Clostridium difficile* strains isolated from an outbreak of antibiotic-associated colitis and diarrhoea. J Med Microbiol 17 : 317-324

35. Price AB, Davies DR (1977) Pseudomembranous colitis. J Clin Pathol 30 : 1-12

36. Rifkin GD, Fekety FR, Silva J Jr (1977) Antibiotic-induced colitis : implication of a toxin neutralized by *Clostridium sordellii* antitoxin. Lancet II : 1103-1107

37. Rolfe RD, Finegold SM (1983) Intestinal β lactamase activity in ampicillin-induced, *C. difficile* associated ileocolitis. J Infect Dis 147 : 227-235

38. Schwan A, Sjolin S, Trottestam U, Aronsson B (1984) Relapsing *Clostridium difficile* enterocolitis cured by rectal infusion of normal faeces. Scand J Infect Dis 16 : 211-215

39. Sjoovall J, Huitefeldt B, Magni L, Nordce I (1986) Effect of β lactam prodrugs on human intestinal microflora. Scand J Infect Dis 49 (Suppl) : 73-84

40. Steinberg JP, Beckerdite ME, Westernfelder GO (1987) Plamid profiles of *Clostri-*

dium difficile isolates from patients with antibiotic-associated colitis in two community hospitals. J Infect Dis 156 : 1036-1038

41. Tabaqchali S, O'Farrell S, Holland D, Silman R (1986) Method for the typing of *Clostridium difficile* based on polyacrylamide gel electrophoresis of (35 S) methionine-labeled proteins. J Clin Microbiol 23 : 197-198

42. Talbot RW, Walker RC, Beart RW Jr (1986) Changing epidemiology, diagnosis, and treatment of *Clostridium difficile* toxin-associated colitis. Br J Surg 73 : 457-460

43. Taylor NS, Thorne GM, Bartlett JG (1981) Comparison of two toxins produced by *Clostridium difficile*. Infect Immun 34 : 1036-1043

44. Toothaker RD, Elmer GW (1984) Prevention of clindamycin-induced mortality in hamsters by *Saccharomyces boulardii*. Antimicrob Agents Chemother 26 : 552-556

45. Triadafilopoulos G, Pothoulakis C, O'Brien MJ, La Mont TJ (1987) Differential effects of *Clostridium difficile* toxins A and B on rabbit ileum. Gastroenterology 93 : 273-279

46. Walker RC, Ruane PJ, Rosenblatt JE (1986) Comparison of culture, cytotoxicity assays, and enzyme-linked immunosorbent assay for toxin A and toxin B in the diagnosis of *Clostridium difficile*-related enteric disease. Diagn Microbiol Infect Dis 5 : 61-69

47. Wilson KH, Sheagren JN (1983) Antagonism of toxigenic *Clostridium difficile* by nontoxigenic *C. difficile*. J Infect Dis 147 : 733-736

Addentum. An important paper appeared in the January 26th 1989 issue of the N Engl J Med 320 : 204-210, on nosocomial acquisition of *C. difficile* infection. In this article, the stools of a large number of hospitalized patients admitted in a general medical ward were studied, together with their hospital personel (hands) and room environment, and isolated *C. difficile* strains were typed by immunoblotting. Seven per cent of patients had positive culture at admission, while 21 per cent acquired the bacterium early or late during their hospitalization. Patient-to-patient transmission of *C. difficile* was evidenced by time-space clustering of incident cases, with identical immunoblot types, and by significantly more frequent and earlier acquisition of the bacterium among patients exposed to rommates with positive culture. Fifty nine per cent of hospital personel caring for patients with *C. difficile*-positive cultures had the bacterium on their hands, and rooms occupied either by symptomatic (49 per cent) or asymptomatic (29 per cent) patients harbouring *C. difficile* became frequently contaminated. Thus, infection with *C. difficile* appears to be frequently transmitted from patients to patients in hospital wards, via the hands of hospital personel and room contamination. Interestingly, the outcomes of nosocomial acquisition of *C. difficile* were asymptomatic carrier state in 14.2 per patients per 100 admissions, antibiotic-induced diarrhea without colitis in 7.1 per cent, and diarrhea not induced by antibiotics in 1,4 p.cent, whereas the corresponding figures for the 29 patients with non incident nosocomial or community-acquired *C. difficile* infection were 59 and 28 per cent, including 3 non specific colitis and 1 PMC. At discharge, often to an extended care facility, 82 per cent of patients with incident cases of *C. difficile* infection still carried the bacterium.

Bacterial overgrowth in children
with severe gastrointestinal disorders

P Chapoy

The intestinal ecosystem (IES) is built upon a reservoir of a hundred billion (10^{14}) bacteria. It plays a very important role in the physiology of the gastrointestinal tract, not only because of its metabolic activity, comparable to that of the liver, but also as a protective screen interacting with the gut mucosa. Its great vulnerability is even more evident in newborns unprovided with mucosal antibody coating and intestinal microflora.

After birth the gastrointestinal tract is continually exposed to potentially harmful external agents : infective, toxic and antigenic.

Intestinal bacterial colonization is one of the rapid physiologic adjustments of early postnatal life [35]. The gut is a sterile organ at birth and newborn babies acquire their fecal floral from their mother. By the second day, coliforms, *Lactobacilli* and *Enterococci* are present. The oxido-reduction potential (Eh) may play an important role in the mechanism of colonization. The growth and fermentation of *E. coli* and *Enterococci* resulting in a fall of Eh to negative values will permit anaerobic bacteria to develop by the third day of the life [22]. Diet, not surprisingly, has a profound effect on intestinal flora, particularly in early life. *Bacteroïdes* become the dominant group in formula-fed babies and *Bifidobacterium* in the breast-fed. Breast-feeding also influences other components of the intestinal microflora : a lack of *E. coli* containing K antigen, decreased number of *Klebsiellae* and *Enterobacteriacae* [26].

The establishment of a functionally active intestinal flora proceeds by stages (short-chain fatty acid production comes first, bilirubin conversion and mucin degradation next, then fecal trypsic activity), and is slower in breast-fed children [33].

The indigenous microflora has an important control mechanism, helping to prevent bacterial overgrowth (coliforms in vitro divide every 20 min as opposed to every 6 to 24 h in the gut) and to confine these microorganisms to the gut lumen. Disequilibrium of the IES produced by immunosuppressors and antibiotics helps to promote bacterial translocation to the mesenteric nodes [6]. The upper limit of the bacterial population in upper intestinal secretions is around 10^3 to 10^4 CFU*/ml and Gram-positives and yeasts are the dominant groups. Distally, the total microbial population increases (10^8 to 10^9 CFU/ml) and anaerobes predominate.

It has been statistically calculated that more than 400 bacterial species constitute the human intestinal microbial ecosystem and any disturbance of this ecosystem will result in a loss of diversity of its members [5].

There are numerous, very complex interactions between microorganisms within the gut lumen and between these microorganisms and the human host :

---

* CFU : colony-forming unit

The indigenous bacteria enhance trophicity, local immunity and disaccharidase activity of the intestinal mucosa. They also interact to promote or prevent intestinal bacterial overgrowth (IBOG) by several mechanisms : competition for substrate, production of antibiotic-like substances such as bacteriocins [8] or byproducts such as short-chain fatty acids.

The major host defense mechanisms against bacterial overgrowth are, according to Freter [19], gastric acid secretion and mucosal peristaltic activity : although the upper gastrointestinal tract of patients with pernicious anemia suffering from achlorhydria is often colonized by fecal type bacteria, mucosal changes in the duodenum are mild, absorption mechanisms rarely affected and the incidence of *Campylobacter* gastritis, according to recent studies presented at the XIII international Congress of Gastroenterology (Rome, 1988) is low ; the interdigestive motor complex appears to be the « house-keeper » of the gut [41].

Mucus secretion and mucosal antibody coating, blocking enterotoxins and preventing bacterial adhesion, are also efficient intestinal barriers.

Pathogenesis

Over recent years it has become apparent that intestinal bacterial overgrowth (IBOG) can occur without a demonstrable anatomical abnormality such as a blind loop and the term « contamined small bowel syndrome » has been proposed. As pointed out by Drude [18], the most common causes of overgrowth are stasis and intestinal barriers defects due to host vulnerability, related to age (immune immaturity under 6 months of age) or underlying disease. The principal stages of IBOG are summarized in Fig. 1 and we will here describe the main related mechanisms of malabsorption.

Steatorrhea : several causes may be responsible for this

Failure of micellar solubilization. Recent data [28] have shown that purely aerobic IBOG as much as anaerobes, may alter duodenal bile-acid composition by hydrolysis and deconjugation. Also, anaerobes such as *Bacteroïdes*, anaerobic *Lactobacilli* and *Clostridia*, have fermentation reactions that enable them to produce volatile fatty acids from unabsorbed longchain fatty acids, contributing to steatorrhea [23].

Functional and morphological damage to the epithelium. This injury may manifest itself in any degree of severity, ranging from depression of small bowel brush border enzymes to complete loss of intestinal villi [1]. Recent work has demonstrated light and electron microscopic morphological and functional damage to the epithelium in both the experimental and clinical blind-loop syndrome [21]. The lesion may be patchy in distribution and characterized by blunting and broadening of villi and increase in the number of mononuclear cells. It is likely that the blood-loss observed in the experimental blind-loop

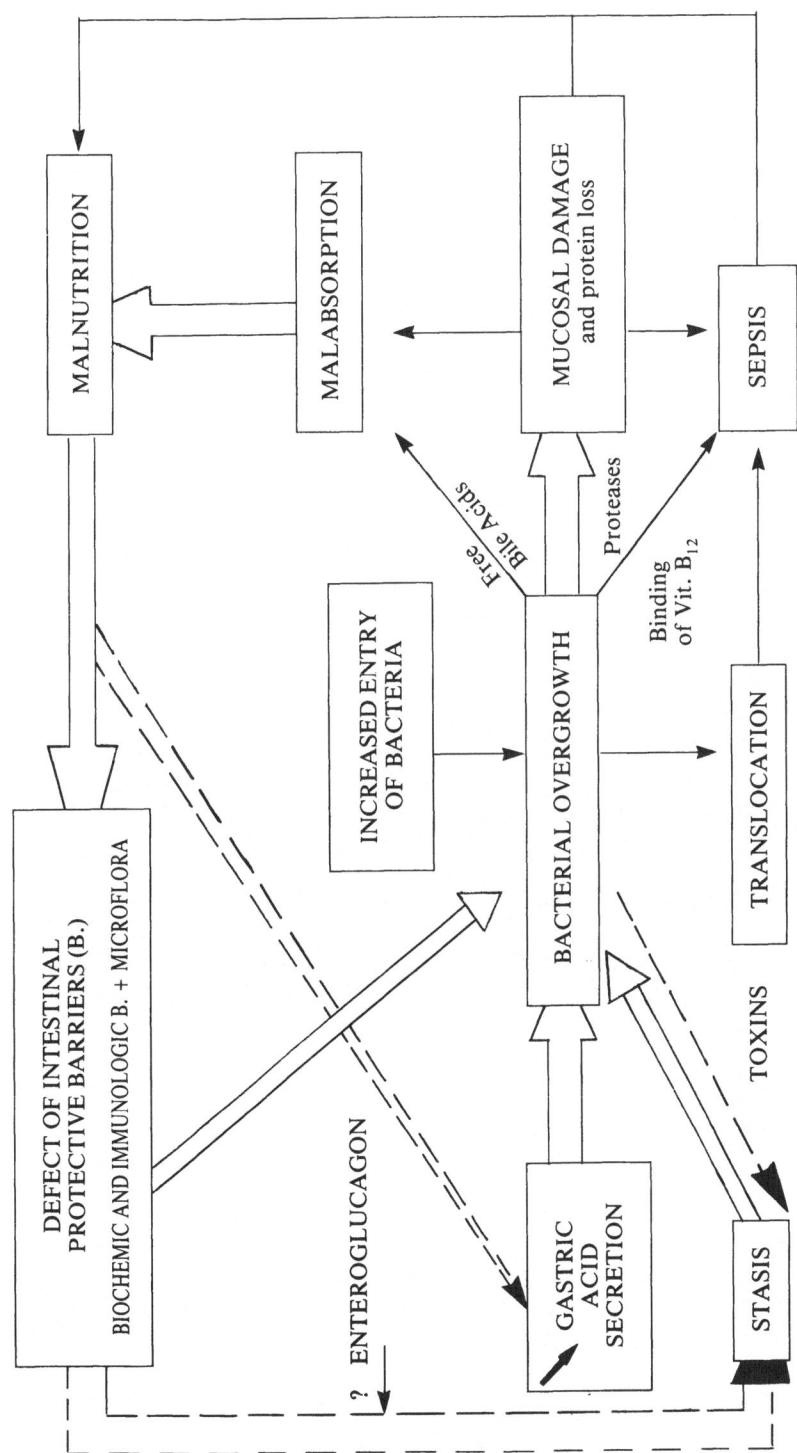

Fig. 1. Schematic representation of vicious cycle of intestinal bacterial overgrowth

syndrome [21] is due to a mucosal lesion by elaboration of enterotoxic byproducts (short-chain fatty acids, free bile acids).

Bacteria also interfer with activity of pancreatic lipase and enterokinase [2],
which accounts for the steatorrhea and protein malabsorption.

Carbohydrates and protein malabsorption

Carbohydrate malabsorption. Anaerobic and facultative anaerobic species involved in the blind-loop syndrome contain proteases capable of removing components of the intestinal surface membrane and may be responsible for disaccharidase deficiency [27]. The possible deleterious effect of unconjugated bile
acids with reduction of fat in the epithelial cells is documented on electron
microscopy. Three bacterial species exhibit a maltase-releasing capacity [27] :
*Bacteroïdes fragilis, Clostridium perfringens, Streptococcus faecalis. Bacteroï-
des* produces protease capable of destroying human brush border disaccharidases at the site of the apolar anchor which binds enzymes to the surface membrane. Increased fermentation will produce organic acids (butyric, propionic,
etc.) which increase osmotic load and impair malabsorption of water and
sodium [17].

Protein malnutrition. Protein malnutrition is multifactorial [4] : catabolism
of ingested protein by the gut flora ; increased enteric loss of endogenous protein from mucosal damage ; increased catabolism of endogenous protein within
the gut lumen by bacterial deaminases ; decreased mucosal transport of dietary aminoacids and peptides [37]. Hypoproteinemia, common in adults, is rare
in children : conversely, osmotic diarrhea due to carbohydrate malabsorption
is often seen in children [2]

Water and electrolyte malabsorption

The two main components of electrolyte depletion are : osmotic diarrhea due
to carbohydrate malabsorption and secretory diarrhea due to bile acid malabsorption and enterotoxins.

Vitamin depletion

Vitamin B_{12} malabsorption. Anaerobic bacteria seem especially able to complete for intrinsic factor (IF) B_{12} and may do so by detaching B_{12} from IF [4].
Vitamin B_{12} is converted into physiologically inactive cyanocobalamine ana-
logs (cobamides). It has been suggested that production of cobamides *de novo*
and from ingested vitamin B_{12} bound to intrinsic factor causes significant
loss [11].

Folic acid deficiency is uncommon as folate levels are often elevated, probably because of microbial synthesis of folic acid-like material in the gut lumen.

Other consequences of bacterial overgrowth

a) *Translocation, which may lead to endogenous septicemia.* b) *Increased mucosal permeability to protein with the risk of food protein sensitivity* (milk protein intolerance).

Diagnosis of the contaminated small bowel syndrome

Clinical manifestations vary greatly according to the nature of the small bowel abnormality causing IBOG. Patients with stricture or a surgically-constructed blind pouch of small intestine may note abdominal discomfort, bloating and crampy periumbilical pain before diarrhea. When patients have strictures or fistulas caused by Crohn's disease, or have hypomotility, the clinical features of the primary disease may be completely overshadowed. The ways in which dysmotility usually becomes manifest include ileus, enterocolitis or septicemia.

Whatever the cause of IBOG, weight-loss associated with clinically apparent steatorrhea is observed in one-third of the patients. Anemia, osteomalacia, hypoproteinemia and vitamin K deficiency may also be observed. Diagnostic tools are listed in Table 1.

Table 1. Laboratory aid in diagnostic of intestinal bacterial overgrowth syndrome

A	Breath hydrogen analysis : after lactulose or glucose load (2 g/kg)	
B	Duodenal intubation :	Differential quantitative study of flora ($> 10^8$CFU) Deconjugated bile acids level /
C	Stool analysis :	Steatorrhea / — Trypsic activity / Stool cultures (proteolytic anaerobic strains) /
D	Plasma vitamins :	B12 \ Folic Acid / Carotene \ (< 40 μg/dl)
E	Urinary indican : /	

/ increase   \decrease

Diagnosis of bacterial overgrowth

There are several laboratory indicators of IBOG.

Indirect tests. These are based on the metabolic activity of enteric bacteria :
- urinary indican level : the bacteria hydrolyze the side-chain of tryptophan to form indican, which can reflect IBOG :
- duodenal unconjugated bile-acid level : samples obtained by duodenal intubation are screened for free bile-acids by thin layer chromatography ;
- stool trypsic activity is reduced in infants with IBOG [7] ;
- breath tests :
• Radioactive CO_2 test : after administration of C^{14} — labeled glychocholic

acid, the 24 h CO_2 excretion is considerably elevated in IBOG ; but the test is not specific and such results can be obtained in ileal dysfunction alone [4].

• H_2 breath test : human cells do not produce hydrogen, and the extremely small amounts excreted via the lungs in healthy individuals comes from the fermentation of carbohydrates by bacteria in the colon. The hydrogen breath test uses this fact to detect, either the presence of bacteria in the small intestine which metabolize ingested sugars, or defective hydrolysis of a disaccharide by a disaccharidase.

After administration of glucose (2 g/kg) or lactulose (2 ml/kg), hydrogen is measured by a gaschromatograph in expiratory flow as schown in Fig. 2 and expressed as particles per million (PPM) [15].

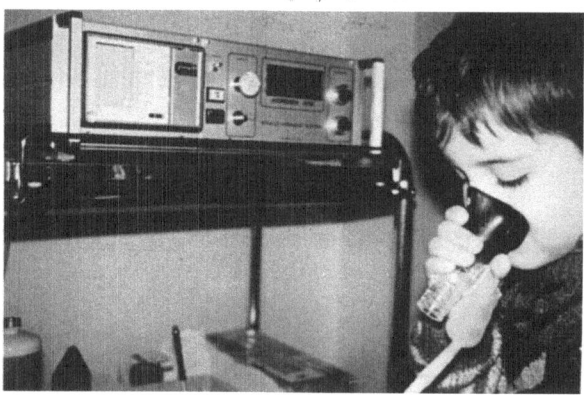

Fig. 2. Exhaled hydrogen monitor system with a Digby-Leigh device to collect expiratory air flow is used to screen intestinal bacterial overgrowth and disaccharidase deficiency

In patients with IBOG the early increase in expired hydrogen is due to fermentation of lactose in the small intestine rather than in the colon [36].

Direct tests. Until recently, the most reliable method of diagnosis of bacterial overgrowth was the detection of numbers of bacteria in samples obtained by duodenal intubation. However this is not a practical test. Differential quantitative analysis of fecal flora (Fig. 3) according to the technique described by Romond [37] seems more useful to the clinician. The samples of feces obtained by rectal swab, and transported in anaerobic Marion scientific culture tubes, are cultured at decimal dilution with Ringer on Columbia media at 37°C. The number of colonies is estimated at day 5 and identification of anaerobes made by morphologic and biochemical analysis. In infants with stasis after surgery, *Pseudomonas* and *Enterobacteriaceae* are isolated significantly more often than in controls, with a major risk of sepsis [21]. IBOG is suspected when CFU/ml of secretion, of both anaerobes and aerobes, is over 10^6 in the duodenum and over 10^9 in the feces. A reasonable estimation of the diversity of human intestinal flora can be obtained by means of Baquero's computerized image-processing analysis [5] : 40 different bacterial morphotypes are recognized from color prints obtained by microphotography of Gram-stained smears

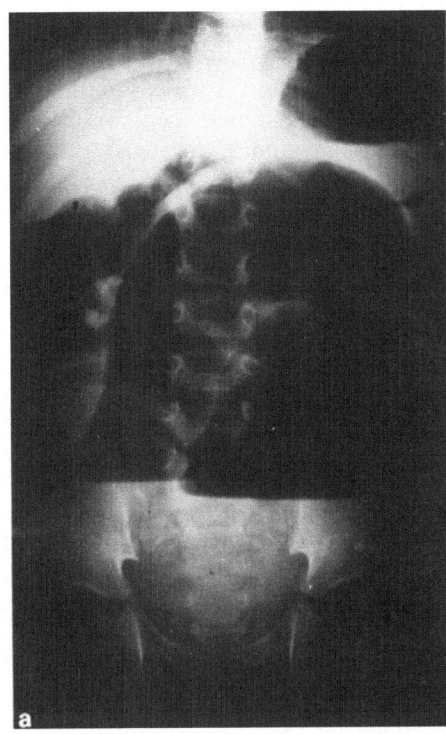

Fig. 3 a, b. Patient with total Hirschsprung's disease and chronic intestinal bacterial overgrowth. **a** Plain film of the abdomen showing chronic ileus ; **b** differential quantitative analysis of fecal flora showing bacterial overgrowth in the same patient

Bacterial strains	Patient	Normal values
Escherochia coli	9.10^6	$10^6 - 10^9$
Streptococcus D	10^4	$10^4 - 10^8$
Anaerobes	12.10^{10*}	$10^6 - 10^{10}$
Enterobacteriaceae :		
Proteus mirabilis	7.10^{6*}	$< 10^7$
Pseudomonas aeruginosa	25.10^{7*}	$< 10^7$
Staphylococcus	10^1	$< 10^7$
Clostridium difficile	16.10^5	$< 10^6$

b

* Bacterial overgrowth

of feces. On average, each individual harbors 15.5 morphotypes and the morphologic diversity index (MDI) = 15.5/40 = 0.38.

MDI is based on several criteria : Gram reaction, regularity of staining, form, length, width and colonies of bacteria.

MDI decreases sharply with broad-spectrum antibiotics and continuous enteral feeding ; it correlates with resistance to colonization and the danger of enteric infection increases when MDI decreases.

Diagnosis of malabsorption

As malabsorption is usually multifactorial, several tests should be performed.
Stool analysis should include : Steatorrhea and creatorrhea ; α 1 antitrypsin
intestinal clearance, to assess protein-losing enteropathy.
Small bowel biopsy (under endoscopic vision) : to document intestinal villous
atrophy.
Serum vitamin levels : carotene and folic acid levels are usually increased ; vitamin B_{12} level is usually decreased and associated with megaloblastic anemia.

Pathologic states associated with IBOG

Bacterial contamination of the upper intestinal secretion can occur in a wide
variety of clinical situations which can be classified according to the host condition : structural defect of the gastrointestinal tract, congenital or acquired ;
functional defect such as dysmotility, hemodynamic failure, immunodeficiency ;
and degree of enteropathogenicity of microbial agents, whether or not selected by antibiotic administration.

This classification is arbitrary and there can be considerable overlap between underlying causes and pathogenic mechanisms. A suggested classification
is given in Table 2.

Post-abdominal surgery IBOG

Patients suffering from post-surgical stasis have significantly high levels of *Pseudomonas* and *Enterobacteriaceae* in the duodenal aspirate [20]. Surgical resection
of the ileum and ileocaecal valve is likely to cause multiple early and late pos-
toperative complications related to stasis and IBOG. Surgical removal of the
terminal ileum also contributes to diarrhea by bile acid-induced chemical colitis and to steatorrhea by decreased bile acid pool. Blind pouches of the small
intestine formed surgically by creation of an end-to-side enteroanastomosis,
frequently produce IBOG [17].

Intestinal dysmotility

Prognosis of congenital or acquired pseudo-obstruction syndromes, with or
without myenteric plexus defect and aganglionosis (Hirschsprung's disease),
depends on the risk of enterocolitis and endogenous sepsis by IBOG mainly
of proteolytic and anaerobic strains.

Malnutrition

Studies of jejunal aspirates of malnourished Australian aborigines and Guate-
malian children show that they have significant IBOG [22]. Also, the oropharyn-

Table 2. Disorders causing intestinal bacterial overgrowth

---

1) *Post-Abdominal surgery* (Complications)
— Post-gastrectomy : achlorhydria ; afferent loop syndrome (Billroth II)
— Post-enteroenterostomy : gastrojejunal (or colic) fistula ; partial obstruction (adhesions) ; blind-loop syndrome ; short gut syndrome (with removal of ileocaecal valve)

2) *Structural defects :*
— *Congenital :*
• Partial or total obstruction : intestinal atresia : diaphragm — bands ; volvulus and malrotation
• Other : duplication — cyst — diverticula
— *Acquired :*
• Crohn's disease ; tuberculosis ; radiation enteritis : stricture — fistula — incompetent ileocaecal valve

3) *Motility disturbances :*
— *Congenital :* Aganglionosis — Pseudoobstruction syndrome
— *Acquired :* Scleroderma ; antral dysmotility ; diabetic enteropathy ; degeneration myenteric plexus

4) *Immunodeficiency :*
— *Congenital :* severe combined immunodeficiency ; hypogammaglobulinemia
— *Acquired :* malnutrition — AIDS

5) *Circulatory failure :* portal hypertension — Necrotizing enterocolitis

6) *Infection :*
— *Viral :* intractable diarrhea
— *Bacterial :* tropical sprue — Postantibiotic and post enteral-feeding diarrhea
— *Parasitic :* Giardiasis

---

geal secretions from undernourished children living in contaminated environments have high levels of fecal organisms and enteropathogens. Impaired gastric acid secretion has also been observed in association with chronic atrophic gastritis [22]. Malnutrition itself causes histologic abnormalities in the upper intestinal mucosa and alterations to its immunologic defenses [25].

Immunodeficiency syndrome

Gastrointestinal disease in children with AIDS may mimic Whipple's disease. It may be due to idiopathic villous atrophy and bacterial or opportunistic infection *(Cytomegalovirus, Cryptosporidium, Candida)*. A case of pseudomembranous necrotizing jejunitis associated with overgrowth of *Klebsiella pneumoniae* has been reported by Loughlin [30].

Necrotizing enterocolitis of prematures

In premature babies, *Klebsiella* are a dominant species and *Lactobacilli* not well-established, so that resistance to bacterial colonization is weak. Necrotizing enterocolitis is another disease of early infancy associated with significant alterations to the gastrointestinal microflora : *Bacteroïdes* and *Lactobacilli* are rarely isolated, as opposed to *Clostridium perfringens.* Colonization with *C. perfringens* in the absence of protective microflora may be crucial in the pathogenesis of necrotizing enterocolitis [7] ; this could be precipitated by neonatal intestinal ischemia, or inappropriate hypertonic artificial feeding [24].

Continuous enteral feeding-induced diarrhea

Diarrhea is still the most common complication of constant rate enteral feeding [12]. It can be attributed to IBOG by several mechanisms : gastric acid exclusion by duodenal intubation [13] ; decreased peristaltic activity [34] ; feeding solution contamination by enteroinvasive bacteria : *E. coli, Klebsiella, Pseudomonas, Acinetobacter, Enterobacter, Staphylococcus, Candida,* during preparation or storage. This explains the need for careful aseptic preparation and continuous refrigeration during infusion. The type of solution can also be incriminated : carbohydrate concentration, osmolality, and associated therapy (H2 antagonists, antibiotics, antiinflammatory agents). However it does not seem to be related to the diet itself [3] : elemental diet does not reduce intestinal microflora but protein hydrolysate may enhance IBOG [34]. This explains why breast-milk feeding may be of interest because of its antiinfectious and prokinetic properties.

Post-gastroenteritis dysbiosis

Viral gastroenteritis. Alteration in the composition of gastrointestinal microflora frequently accompanies acute viral or bacterial diarrheal illnesses, characterized by an increased number of *Fusobacteria* and *Clostridia* [38] with increase of total concentration of bacteria in the duodenum up to 100 times normal values [40]. The concentration of *Bacteroïdes* is decreased in the feces. Such changes may reduce resistance to bacterial colonization and be responsible for intractable diarrhea.

Bacterial gastroenteritis. A major cause of diarrheal illness in children of developing countries and in travelers is enterotoxigenic *E. coli*, rather than non-enterotoxigenic strains [28]. It is frequently associated with an increased concentration in the jejunal secretion, of *Klebsiella, Proteus* and *Pseudomonas.* The scope for toxin production by intestinal bacteria is much more extensive than suspected in the past and certainly not confined to *E. coli* ; *Aeromonas hydrophila* is an illustrative example [24].

Tropical sprue. The upper gut microflora in the tropics is unbalanced because of enterotoxigenic strain contamination, malnutrition [26] and immunodeficiency [30]. The malabsorption syndrome is consecutive to colonization of the proximal bowel with enterotoxigenic coliform bacteria *(Klebsiella, Enterabacter, E. coli)* and a decreased number of anaerobes. Enteroglucagon secretion resulting from mucosal cell injury promotes bacterial strains and enhances IBOG.

Antibiotic-induced diarrhea. The administration of broad-spectrum antibiotics may cause diarrhea. They are responsible for suppression of sensitive strains, which may have a microbial barrier effect, and replacement by resistant organisms [10]. This leads to IBOG and possibly endogenous sepsis. Antibiotic treatment of the mother has an important effect on the newborn's flora if the treatment is given before delivery [9].

Pseudomembranous colitis is the major complication and is due to *Clostridium difficile* cytotoxin. The same *Clostridium* producing toxin is found in significant concentrations in the large microflora of healthy infants [32] ; the carriage rate falls sharply after the first year of life [41]. The protection of infants remains a puzzling question : a diet effect as shown in mice by Mahe [31] might be an explanation.

Intractable diarrhea. Intractable diarrhea, according to Rossi [39], is not a specific diarrheal disease but a group of illnesses in which diarrhea is not readily explained but most probably infectious (viruses, toxigenic bacteria) in nature, although, the cause is not always demonstrated. Intestinal biopsies have shown variable degrees of histologic damage with villous shortening, blunting and inflammatory infiltration of the lamina propria. Apart from specific enteropathogenic bacteria, infants with protacted diarrhea have an abnormally profuse, mixed microbial flora in the upper gut [14]. This could be related to a significant adherence facilitation of the enterocytes of these patients to certain strains (*E. coli* 01 K1 H7). Microbial degradation of bile salt is a major factor of malabsorption in this syndrome.

Treatment

The three main approaches to treatment of IBOG are : bacterial decontamination, nutritional support and surgery.

Bacterial decontamination

This can be managed with 2 types of drugs.

Antibiotics. Attention should be paid to the side-effects of broad-spectrum antibiotics and the need for probiotics in prophylaxis of this syndrome.

Although it would seem desirable to select an appropriate antibiotic based on the result of cultures of small intestinal secretions, this approach is rarely feasible in practice. In the near future, MDI [5] calculated by computerized technique will probably help in screening high-risk patients. Anaerobic coverage is always required, and antifungal agents if yeast infection is documented. *Tetracyclines* should not be used before 6 years of age to prevent dental complications ; *Ampicillin* et *Lincomycin* should be disregarded because of their implication in pseudomembranous colitis. *Neomycin* is inefficient when used alone.

In fact, sequential intestinal decontamination with alternated *Metronidazole* (30 mg/kg), *Aminoglycosides* (6 mg/kg) and sometimes *Vancomycin* (25 mg/kg), *Erythromycin* (50 mg/kg) or *Trimethoprim sulfamethoxazole* (30 mg/kg) seems to be the most efficient approach.

Antitoxin agents. Cholestyramin (50 mg/kg) is an ion-exchange resin and bile sequestrant ; it may prevent the stimulating effect on colonic secretion and hyperoxaluria.

Probiotics. Probiotics are living microorganisms, and not their fermentation products, which act as bioregulators of intestinal flora through their antagonistically acting metabolic products. They are able to adhere to the intestinal wall to form biofilm protective against pathogens. *Saccharomyces boulardii* (50 mg/kg), a mesophilic non-pathogenic yeast, known for its antidiarrheal and antitoxinic effect is probably the most efficient. It can be used prophylactically but also curatively because of its microbial antagonistic effect [16], particularly against *Candida* and *Pseudomonas*.

Nutritional support

It is clear that IBOG may lead to a wide spectrum of absorption defects that can cause multiple nutritional deficiencies. Nutritional rehabilitation should be carefully planned. This includes :
- vitamin and trace metal supplementation : monthly injection of vitamin B_{12} ; vitamin D and vitamin K adminstration when recommended ;
- medium-chain triglycerides and sometimes protein hydrolysates and glucose polymers administered by constant-rate enteral feeding, may be necessary to reverse nitrogen balance.

Surgery

Surgery should be planned on patients already prepared by nutritional support and intestinal decontamination. Each situation should be carefully assessed according to the likely benefits and long-term side-effects from the operative procedure planned. The recurrent nature of Crohn's disease, for instance, amenable to surgical repair should be considered.

References

1. Ament M, Shimoda S, Saunders D, Rubin C (1972) Pathogenesis of steatorrhea in three cases of small intestinal stasis syndrome. Gastroenterology 63 : 728-747
2. Anderson C, Burke V (1975) Bile and bacteria 12 : 397-410. In : Pediatric gastroenterology (text book). Blackwell Scientific Publications, Oxford

3. Axelsson C, Justsesen T (1977) Studies of the duodenal and fecal flora in gastrointestinal disorders during treatment with an elemental diet. Gastroenterology 72 : 397-401

4. Banwell G (1981) Small intestinal bacterial overgrowth syndrome. Gastroenterology 80 : 834-845

5. Baquero F, Fernandez-Jorge A, Vicente MF, Alos JL, Reig M (1988) Diversity analysis of the human intestinal flora : a simple method based on bacterial morphotypes. Microb Ecol Health Dis 1 : 1-8

6. Berg RD (1980) Mechanisms confining indigenous bacteria to the gastrointestinal tract Am J Clin Nutr 33 : 472-484

7. Blakey JL, Lubitz L, Campbellnt, Gillam GL, Bishop RF, Barnes GL (1988) Enteric colonization in sporadic neonatal necrotizing enterocolitis. J Pediatr Gastroenterol Nutr 7 : 559-567

8. Booth SJ, Johnson JL, Wilkins TD (1977) Bacteriocin production by strains of Bacteroides isolated from human feces and the role of these strains in the bacterial ecology of the colon. Antimicrob Agents Chemother 11 : 718-724

9. Borderon JC, Bernar JC, Vergnand R, Gold F (1980) Effet de l'antibiothérapie de la mère sur la colonisation du nouveau-né par les entérobactéries. Arch Fr Pediatr 37 : 371-376

10. Bourrillon A, Lambert-Zechovsky N, Beaufils F, Lejeune C, Bingen E, Blum C, Mathieu H (1978) Antibiothérapie et pullulation microbienne intestinale et risque infectieux chez l'enfant. Arch Fr Pediatr 35 : 23-37

11. Brandt LJ, Bernstein LH, Wagle A (1977) Production of vitamin B12 analogues in patients with small bowel bacterial overgrowth. Ann Inter Med 87 : 546-551

12. Cano N, Di Costanzo J, Chapoy P, Martin J, Richieri JP (1987) Nutrition enterale de l'adulte. Encycl Med Chir Paris — Estomac-Intestin, 9110 A 10 : 12

13. Chalacombe DN (1974) Bacterial microflora in infants receiving naso-jejunal tube feeding. J Pediatr 85 : 113

14. Chalacombe DN, Richardson JM, Rowed B, Anderson CM (1974) Bacterial microflora of the upper gastroinstestinal tract in infants with protracted diarrhea. Arch Dis Child 49 : 270-277

15. Chapoy P (1988) Ecosystème intestinal de l'enfant : I et II. Concours Médical 120 : 1606-1610 and 1699-1707

16. Chapoy P (1986) A propos du mode d'action intestinal de Saccharomyces boulardii (lettre). Gastroenterol Clin Biol 10 : 860-861

17. Donaldson R (1978) The blind loop syndrome. II : Gastrointestinal diseases — Shleisinger-Fordtran. Saunders, Philadelphia 63 : 1094-1103

18. Drude R, Chesley Hines JR (1980) The pathophysiology of intestinal bacterial overgrowth syndrome. Arch Intern Med 140 : 1349-1352

19. Freter R (1974) Interactions between mechanisms controlling the intestinal microflora. Am J Clin Nutr 27 : 1409-1416

20. Ghnassia JC, Ricour C, Duhamel JF, Nihoul-Fekete C, Veron M (1978) Flore bactérienne duodénale de l'enfant au cours des stases post-chirurgicales. Arch Fr Pediatr 35 : 854-862

21. Gianella R, Toskes P (1976) Gastrointestinal bleeding and iron absorption in the experimental blind loop syndrome. Am J Clin Nutr 29 : 754-757

22. Gracey M, Stone DE (1972) Small intestinal microflora in australian aboriginal children with chronic diarrhea. Austr NZ J Med 3 : 215-219

23. Gracey M (1979) The contamined small bowel syndrome : pathogenesis, diagnosis and treatment. Am J Clin Nutr 32 : 234-243

24. Gracey M (1982) Intestinal microflora and bacterial overgrowth in early life. J Pediatr Gastroenterol Nutr 1 : 13-22

25. Gracey M, Burke V, Robinson J (1982) Aeromonas-associated gastroenteritis. Lancet II : 1304-1306

26. Gracey M (1984) The intestinal microflora in malnutrition an protracted diarrhea in infancy. In : Lebenthal E (Ed) Chronic diarrhea in children. Raven Press, New York, pp 223-237

27. Jonas A, Krishman C, Forstner G (1978) Release of disaccharidases from brush border membranes by extracts of bacteria obtained from intestinal blind loops in rats. Gastroenterology 75 : 791-795

28. Kocoshis SA, Schletewitz K, Livelace G, Laine RA (1987) Duodenal bile acids among children : keto derivatives and aerobic small bowel bacteria overgrowth. J Pediatr Gastroenterol Nutr 6 : 686-696

29. Levine MM (1987) Escherichia coli that cause diarrhea : enterotoxigenic, enteropathogenic, enteroinvasive, enterohemorrhagic, and enteroadherent. J Infect Dis 155 : 377-389

30. Mc Louchlin LC, Nor KS, Joshi VV, Oleske JM, Connor EM (1987) Severe gastrointestinal involvment in children with the acquired immunodeficiency syndrome. J Pediatr Gastroenterol Nutr 6 : 517-524

31. Mahe S, Corthier G, Dubos F (1987) Effect of various diets on toxin production by two strains of Clostridium difficile in gnotobiotic mice. Infect Immun 55 : 1801-1805

32. Merida V, Moerman J, Colaert J, Lemmens P, Vandepitte J (1986) Significance of Clostridium difficile and its cytotoxin in children. Eur J Pediatr 144 : 494-496

33. Midtvedt AC, Carlstedt-Duke B, Norin KF, Saxerholt H, Midtvedt T (1988) Development of five metabolic activities associated with the intestinal microflora of healthy infants. J Pediatr Gastroenterol Nutr 7 : 559-567

34. Navarro J, Lambert-Zechovsky N, Cezard JP (1983) Alimentation et écosystème bactérien intestinal en pathologie digestive pédiatrique. Arch Fr Pediatr 40 : 677-679

35. Neter E, Braun DA (1981) Microbial colonization of the newborn. In : Lebenthal E (Ed) Textbook of gastroenterology and nutrition in infancy. Raven Press, New York, pp 239-246

36. Rhodes JM, Middleton P, Welld (1974) The lactulose hydrogen breath test as a diagnostic test for small bowel bacteria overgrowth. Scand J Gastroenterol 14 : 336

37. Riepe S, Goldstein J, Alpers DH (1980) Effect of secreted Bacteroïdes proteases on human intestinal brush border hydrolases. J Clin Invest 66 : 314-322

38. Romond C, Neut C, Beerens H, Turck D, Fontaine G (1981) Dysmicrobisme anaérobie et diarrhée chez le nourrisson. Rev Inst Pasteur-Lyon 14 : 289-301

39. Rossi TM, Lebenthal E (1981) Intractable diarrhea of infancy. In : Leventhal E (Ed) Textbook of gastroenterology and nutrition in infancy. Raven Press, New York, pp 987-1001

40. Simon G, Gorbach S (1986) The human intestinal microflora. Digest Dis Sci 31 : 147-162

41. Stark PL, Lee A, Parsonage BD (1982) Colonization of the large bowel by Clostridium difficile in healthy infants ; quantitative study. Infect Immun 35 : 895-899

42. Vantrappen G, Janssens J, Hellemans J (1977) The inter digestive motor complex of normal subjects and patients with bacterial overgrowth of the small intestine. J Clin Invest 59 : 1158-1166

Effects of antibiotherapy on microbial intestinal ecosystem (MIE) in newborns (NB) and children (Ch)

Y Aujard, N Lambert-Zechovsky, E Bingen, A Bourrillon, and H Mathieu

The digestive tract is sterile at birth [16, 20, 21]. In a few days bacterial concentrations are nearly those observed in older children and adults. The stability of MIE is remarkable, whatever the age. The organisms forming the intestinal floral continually exert forces to maintain an inter-species equilibrium. The practical effect is the exclusion of non-indigenous organisms, including pathogens [12]. Disturbances of MIE induced by antibiotics lead to pullulation of a bacterial species which can be, by translocation, responsible for an acquired septicemia [10, 18].

We have studied, by a qualitative and quantitative analysis of the fecal flora, the effects of antibiotics on MIE in neonates and children.

This study was conducted within a paediatric hospital environment.

Patients and methods

Ninety-eight neonates and infants were studied. All received antibiotherapy because of a suspected infection : neonatal sepsis, respiratory tract or urinary tract infections.

Antibiotics given were : 1) IV : penicillin G (n = 6) ; ampicillin (n = 10) ; amoxycillin (n = 4) ; amoxycillin and clavulanic acid (n = 6) ; cefotaxime (n = 25) ; cefoperazone (n = 10) ; cefotiam (n = 10) ; mezlocillin (n = 6) ; fosfomycin (n = 7).

2) per os : ampicillin (n = 10) ; amoxycillin (n = 4) ; amoxycillin and clavulanic acid (n = 3) ; erythromycin (n = 5) ; pristinamycin (n = 7) ; colistin (n = 10).

Stools were collected under sterile conditions and were immediaty analyzed in the laboratory by a previously described technique [8] :
— examination by phase-contrast and dark-field microscopy, and Gram staining ;
— 1 g (one gramme) of feces were homogenized with an ultra turrax in 9 ml of sterile water and dilutions from 10^2 to 10^9 were made in sterile water ;
— 0.1 ml of each dilution was cultured in different selective or non-selective media under anaerobic and aerobic conditions ;
— after incubation every bacterial species was counted with a threshold of 10^2 organims/g of stools and identified by the usual biochemical, immunological and culture methods. To obtain a precise epidemiologic identification, we used : 1) the biotype for *Escherichia coli* and *Enterobacter cloacae* ; 2) the biotype and capsular serotype for *Klebsiella* and *Enterobacter aerogenes* ; 3) the

biogroup for *Enterobacter agglomerans* ; 4) the serotype for *Pseudomonas aeruginosa* and 5) the serotype and lysotype for staphylococci. It was thus possible to identify and numerate the bacterial strains in each stool sample. The phenotype of resistance to antibiotics was also established according to the techniques and norms of the Pasteur Institute.

The intestinal flora was studied on admission and then every 3 to 4 days.

Normal lower and upper concentrations of each organism were previously defined [8]. Overgrowth is defined for each strain species as above the upper limits of the study group and abnormal decrease of bacteria as a count below the lower limit of the study group. The absence or disappearance of a bacterial species is defined as a number of bacteria below the threshold of 10^2 organisms/g of feces.

Results

Antibiotics per os (Table 1)

Betalactams, pristinamycin and colistin with total parenteral nutrition caused a decrease or disappearance of sensitive bacteria and an overgrowth of resistant organims.

With all beta-lactams, ampicillin, amoxycillin and amoxycillin combined with clavulanic acid, the effects on the MIE were the same, an overgrowth of *Klebsiella*. In some cases, *E. coli* remained in the intestinal flora and even proliferated.

With pristinamycin, overgrowth of *Klebsiella*, and/or *Proteus, Enterobacter* and *Pseudomonas aeruginosa* was noticed.

When colistin was given with feeding no modification was found in MIE.

Table 1. Antibiotics given orally

	Overgrowth	Decrease or disappearance
— Amoxycillin	Klebsiella	E. coli
— Amoxycillin and Clavulanic acid		
— Erythromycin	none (Enterobacter cloacae)	Enterobacteriaceae Gram + cocci
— Pristinamycin	Klebsiella Enterobacter	Staphylococci
— Colistin		
. fed patients	none	none
. non fed patients	Proteus Staphylococci	Gram-bacilli

Without feeding there was a disappearance or, at least, a decrease of Gram-negative susceptible bacteria, such as *E. coli, Klebsiella, Enterobacter* ; conversely a proliferation of proteus and/or *Staphylococcus aureus* was observed.

With erythromycin there was observed a paradoxical decrease in all Gram-positive susceptible and Gram-negative resistant bacterial species. These effects are explained by the high concentration of erythromycin in the digestive lumen (1-10 mg/g stool), more than 10 times the MIC of resistant bacteria.

Antibiotics given by IV or IM routes (Table 2)

No effect was observed with antibiotics having no intestinal excretion (aminoglycosides, colistin). Penicillin, ampicillin, amoxycillin, with or without clavulanic acid, have the same effects on MIE by IV and by mouth (Fig. 1).

Table 2. Antibiotics given by intravenous or intramuscular routes

	Overgrowth	Decrease or disappearance
— Ampicillin ⎰ — Amoxycillin ⎱	Klebsiella	E. coli
— Amoxycillin and Clavulanic acid	Klebsiella Enterobacter	
— Penicillin G	E. coli Klebsiella	none
— Cefotaxime	Pseudomonas aeruginosa	
— Cefoperazone	Yeasts	Gram + and Gram − bacilli
— Cefotiam	Enterobacteriaceae	
— Mezlocillin	none	Enterobacteriaceae
— Fosfomycin	Enterobacteriaceae	Staphylococci

With cefotaxime, there was a decrease in *Enterobacteriaceae* (*E. coli* ; Streptococcus D) and an overgrowth of *Pseudomonas aeruginosa*. Cefoperazone which has an important (70 %) biliary excretion, induced a decrease of all bacterial species ; in 50 % of cases yeasts (Candida) were selected.

No overgrowth was observed with mezlocillin but a decrease of Enterobacteriacae occurred.

Fosfomycin induced an overgrowth of Enterbacteriaceae, as did cefotiam.

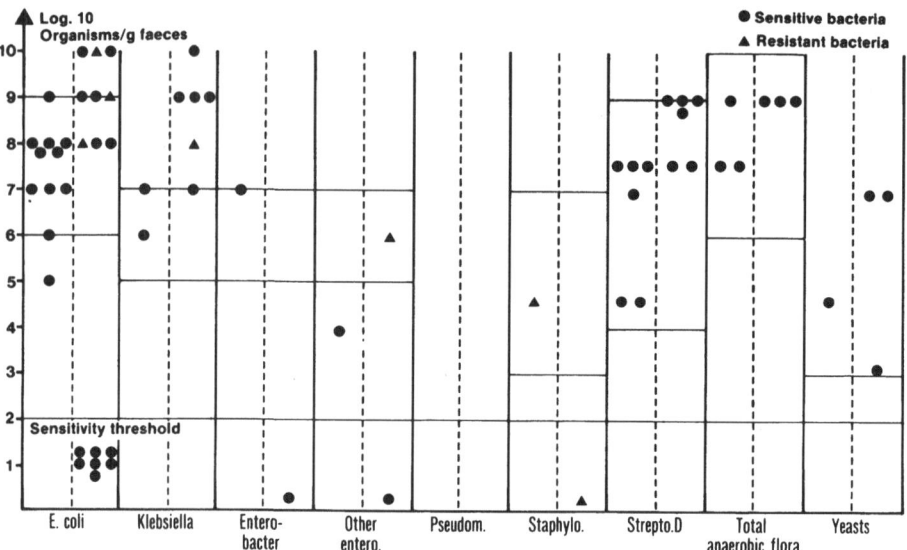

Fig. 1. Effects of amoxycillin-clavulanic acid on fecal flora

Antibiotic-induced pullulation and septicemia by translocation

In 24 newborns presenting a secondary septicemia during antibiotherapy, we have demonstrated the following sequence : MIE normal before antibiotics, overgrowth of a bacterial species during treatment and septicemia with the same pathogen.

All the neonates were treated for a suspected or proved infection with a combination of an aminoglycoside and penicillin G (n = 4), ampicillin (n = 16), cefotaxime (n = 4).

The pathogens responsible for the septicemia were *E. coli* (n = 2), *Enterobacter* (n = 1), *Klebsiella* (n = 1) with penicillin G ; *Klebsiella* (n = 15) and *Enterobacter* (n = 1) with ampicillin ; *Pseudomonas* (n = 4) with cefotaxime. The concentrations of the pullulated organisms in feces ranged between a decrease and disappearance of sensitive species, worsening the MIE unbalance. The mean delay between the onset of antibiotics and septicemia was 8 days (range 1 to 11 days).

Two evolutions of septicemia due to translocation are illustrated by 2 clinical cases :

Case 1 : in a neonate treated for a suspected sepsis by ampicillin and gentamicin, overgrowth of a *Klebsiella* was noticed on day 6 ; on day 10 blood culture was positive for this bacterium. Treatment by cefotaxime and gentamicin cured the systemic infection and reduced *Klebsiella* pullulation in MIE (Fig. 2) ;

Case 2 : a Streptococcus B-infected neonate was treated by ampicillin and gentamicin. On day 3 of treatment, a *Klebsiella* overgrowth was detected, complicated by septicemia. Colistin and gentamicin IV were given on day 4 without effect on the septicemia (blood-cultures positive on day 6). Cefoperazone was used on day 9 and the systemic infection was cured with a dramatic decrease of all the bacterial species in MIE.

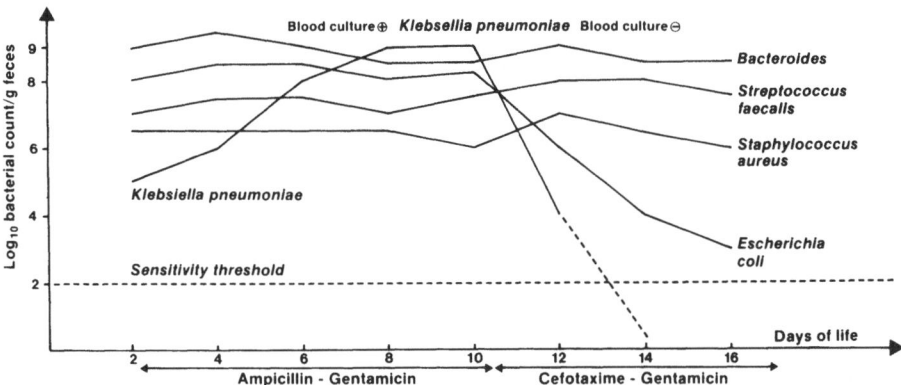

Fig. 2. Fecal flora of a newborn with a *Klebsiella* sepsis

Discussion

After a normal delivery, the neonate's intestinal flora is sterile. At 48 to 72 h of life, the fecal flora contains equilibrated concentrations of bacteria very close to those observed in older children and adults. Each species is very stable, which allows us to define the lower and upper limits of normal concentrations. Chronologic and qualitative sequences of MIE installation depend in part on the mode of delivery, vaginal versus cesarean section, the mode of feeding, human versus cow milk [17, 20], the iron content of milk [19] and the gestational age [21].

MIE equilibrium is disturbed in other pathological circumstances [1, 10] :
— colonization of the mother's vagina by quantitatively or qualitatively abnormal flora can induce massive contamination of the newborn ;
— intestinal stasis leads to an overgrowth of all bacterial species ;
— necrotizing enterocolitis has been related to various factors such as hypoxemia and bacterial pullulation ; it includes *Clostridium* species but also other bacteria, like *Klebsiella*.

However, all the antibiotics given orally or by parenteral route with an intestinal secretion have dramatic effects on MIE. These perturbations are usually a decrease of sensitive species and an overgrowth of resistant species. Each antibiotic induces a selective pullulation : *E. coli* and *Klebsiella* for penicillin G [1, 10] ; *Klebsiella* for ampicillins [1, 10, 14] ; *Pseudomonas* for cefotaxime [9], yeasts for cefoperazone [15].

The major risk of antibiotic-induced bacterial overgrowth is bacterial translocation and bacteremia or septicemia. Usually, overgrowing bacteria are resistant to the initial antibiotic. In some cases it can be noted a discordance with

two explanations : a very high local concentration of the antibiotic (erythromycin) or the presence of β-lactamase in the feces. Other adverse effects of antibiotics on MIE have been described, but were not noted in our study :
— hypoprothrombinemia and platelet dysfunction observed in some adult cases treated by moxalactam [2] ;
— carbohydrate intolerance in term neonates [7].

Thus, the pharmacokinetic properties of an antibiotic are decisive for its choice in the treatment of neonatal sepsis. A poor intestinal secretion — for parenteral antibiotics, and a high intestinal absorption — for oral antibiotics, as can be achieved by ampicillin prodrugs [6], lead to a less serious impact on MIE.

Conclusions

The knowledge of bacterial intestinal translocation as the major source of secondary septicemia has allowed a better use of antibiotics in neonates. In suspected but unconfirmed infections, treatment is stopped on day 3.

If, during antibiotic treatment, secondary sepsis occurs, such a mechanism has to be suspected, and can be confirmed by MIE study. Nevertheless, the knowledge of selected pullulation allows us to start an effective antibiotherapy before the results of blood cultures.

References

1. Aujard Y, Lambert-Zechovsky N, Bourrillon A, Bingen E (1983) Antibiothérapie et écosystème microbien intestinal chez l'enfant. Journées Parisiennes de Pédiatrie. Flammarion, Paris, pp 271-279
2. Bang NV, Tessler SS, Heidenreich RO, Marks CA, Mattler LE (1982) Effects of moxalactam on blood coagulation and platelet function. Rev Inf Dis 4 (Suppl) : 546-554
3. Bennet R, Eriksson M, Nord CE, Zetterström R (1982) Suppression of aerobic and anaerobic flora in newborns receiving parenteral gentamicin and ampicillin. Acta Paediatr Scand 71 : 559-62
4. Bennet R, Eriksson M, Nord CE, Ztterström R (1986) Fecal bacterial microflora of newborn infants during intensive care management and treatment with five antibiotic regimens. Pediatr Infect Dis 5 : 553-559
5. Bennet R, Nord CE (1987) Development of the fecal anaerobic microflora after caesarean section and treatment with antibiotics in newborn infants. Infection 15 : 332-6
6. Bergan T (1986) Pharmacokinetic differentiation and consequences for normal microflora. Scand J Infect Dis 49 (Suppl) : 91-99
7. Bhatia J, Prihoda AR, Richardson J (1986) Parenteral antibiotics and carbohydrate intolerance in term neonates. Am J Dis Child 140 : 111-113
8. Bingen E, Lambert-Zechovsky N (1984) Technical aspects of the quantitative and differential analysis of the microbial intestinal ecosystem. Dev Pharmacol Ther 7 (Suppl) : 134-137

9. Bourrillon A, Brackman D, Boussougant Y, de Paillerets F (1984) Cefotaxime effects on the intestinal flora of the newborn. Dev Pharmacol Ther 7 (Suppl) : 144-149

10. Bourrillon A, Lambert-Zechovsky N, Beaufils F, Lejeune C, Bingen E, Mathieu H (1978) Antibiothérapie et pullulation microbienne intestinale et risque infectieux chez l'enfant. Arch Fr Pediatr 35 : 23-37

11. Finegold SM (1986) Intestinal microbial changes and disease as a result of antimicrobial use. Pediatr Inf Dis 5 (Suppl) : 88-90

12. Hentges DJ (1986) The protective function of the indigenous intestinal flora. Pediatr Inf Dis 5 : 17-20

13. Lambert-Zechovsky N, Bingen E, Bourrillon A, Aujard Y, Mathieu H (1984) Effects of antibiotics on the microbial intestinal ecosystem. Dev Pharmacol Ther 7 (Suppl) : 150-157

14. Lambert-Zechovsky N, Bingen E, Proux MC, Aujard Y, Mathieu H (1984) Influence de l'amoxicilline associé à l'acide clavulanique sur la flore fécale de l'enfant. Pathol Biol 32 : 436-38

15. Lambert-Zechovsky N, Bingen E, Proux MC, Aujard Y, Mathieu H (1984) Influence de la Céfopérazone sur la flore fécale de l'enfant. Pathol Biol 32 : 439-442

16. Lejeune C, Boussougant Y, de Paillerets F (1981) Séquences d'installation de la flore du nouveau-né. Étude par analyse différentielle quantitative. Rev Pediatr 17 : 223-243

17. Lundequist B, Nord CE, Winberg J (1985) The composition of the fecal flora in breast fed and bottle fed infants from birth to eight weeks. Acta Paediatr Scand 74 : 45-51

18. Mathieu H, Lambert-Zechosky N, Bourrillon A, Bingen E, Aujard Y, Beaufils F (1984) Antibiotic therapy and bacterial overgrowth in intestinal microbial ecosystem : a major risk of secondary infections in newborns. Dev Pharmacol Ther 7 (Suppl) : 158-163

19. Mevissen-Verhage EAE, Mercelis JH, Hermsen-Van Ameromgen WCM, de Vos NM, Verhoef J (1985) Effect of iron on neonatal gut flora during the first three months of life. Eur J Clin Microbiol 4 : 273-278

20. Moreau MC, Thomasson M, Ducluzeau R, Raibaud P (1986) Cinétique d'établissement de la microflore digestive chez le nouveau-né humain en fonction de la nature du lait. Reprod Nutr Dev 26 : 745-753

21. Sakata H, Yoshioka H, Fujita K (1985) Development of the intestinal flora in very low birth weight infants compared to normal full term newborns. Eur J Pediatr 144 : 186-190

Antibiotic associated diarrhea : risk factors and reduction of incidence by the yeast, *Saccharomyces boulardii*

CM Surawicz, GW Elmer, P Speelman, LV McFarland, J Chinn, and G Van Belle

Antibiotic associated diarrhea is one of the most frequent adverse effects of antimicrobial agents. Although usually self limiting upon termination of the antibiotic, it may not always be feasible to stop antibiotherapy. For some patients, diarrhea may exacerbate an already serious condition by dehydration and/or by perturbation to the electrolyte balance. Most antibiotic associated diarrhea is thought to be due to an imbalance in the intestinal flora created by destruction or suppression of those components of the intestinal flora that exhibit a barrier effect against the overgrowth of diarrhea causing pathogens [21, 22]. Increases in unabsorbed carbohydrates or altered fatty acids in the gut may also play roles in the etiology [17, 18].

Very few attempts have been made to systematically determine whether exogenous administration of nonpathogenic microbes to replace those microflora components destroyed by antibiotherapy would decrease antibiotic associated diarrhea. A *Lactobacillus* preparation has been reported to reduce ampicillin associated diarrhea [12], yogurt reduced erythromycin induced diarrhea [6], and fecal enemas [3, 19] and a special *Lactobacillus* preparation [13] reduced relapses of pseudomembranous colitis. A yeast would have some inherent advantage as a microbial agent for preventing antibiotic associated diarrhea. *Saccharomyces boulardii* is used in Europe and some other countries to prevent and treat antibiotic associated diarrhea and there is evidence that it is efficacious for this purpose [1]. We chose to evaluate the efficacy of this yeast in preventing diarrhea in a hospital setting where the use of multiple broad spectrum antibiotics was frequent and the incidence of antibiotic associated diarrhea was high (26 %, LV McFarland, unpublished data). As part of the study we also have quantitated risk factors for antibiotic associated diarrhea, including *Clostridium difficile*.

Methods

Consecutive patients at Harborview Medical Center, Seattle, Washington, USA, who received antibiotics were eligible for the study. Patients were not eligible if they had diarrhea within the previous week or within 24 h after enrollment, if they were pregnant, or if the duration of antibiotherapy was less than 3 days. Patients were also excluded if they received vancomycin, metronidazole, antifungal antibiotics, or lactulose. Enrolled patients were randomized in 2 : 1 blocks to *S. boulardii* (two 250 mg capsules) or placebo (indistinguisable from the yeast capsules) taken twice a day. The study drug was taken during the antibiotic therapy and for 2 weeks afterwards. Stool samples were submitted for analyses on entry to the study, on day 5, thereafter about every 10 days and upon completion of the study. Stool consistency and frequency were monitored daily. Diarrhea was defined as three or more loose or watery stools per day for 2 days.

Stools (or in some cases rectal swabs) were cultured for *C. difficile* using standard techniques [10]. To detect low levels of *C. difficile*, prereduced broth media were inoculated and incubated anaerobically [15]. Cytotoxin titres of *C. difficile* positive stools were determined as described [11]. Differences between means were assessed using Student's t test and differences between group proportions using the Chi-square statistic [14]. When the sample size was small, differences were assessed using Fisher's Exact Test.

Relative risks were calculated from incidence density ratios based on days of specific antibiotic use. Incidence density ratios were calculated from the formula : $[N_e/(pt_e) : (N_o/pt_o)]$; where N_e equals the number of patients with incident diarrhea exposed to a specific antibiotic, pt_e' equals the total of the person-days of an antibiotic used by diarrheal and non-diarrheal patients. The denominator (N_o/pt_o) is a common baseline taken from patients whose only antibiotic exposure was to penicillin. Days of exposure after the onset of diarrhea were excluded. Significance of the relative risks was indicated when the 95 % test-based confidence intervals excluded one [14].

Results

The demographic characteristics of the evaluable and unevaluable patients are shown in Table 1. We had a relatively large number of unevaluable patients (138/318) because of our requirement that only patients monitored for 8 or more days be included in the final analysis for diarrhea. This analytical criterion was instituted because antibiotic associated diarrhea takes time to develop (5.5 ± 3.8 days in this study) and if the total population were included in the computations, patients would not have been followed for a sufficient period of time. Other patients were dropped (64/138) for the following rea-

Table 1. Demographics of the study population

Population	N	Age			Sex
		mean	SD	median	% males
Drops	64	50.3	20	51.0	62.5
Excluded					
(< 8d obs)	74	46.4	19	44.0	69.3
Included	180	47.8	20	44.5	68.9
Placebo	64	46.2	19	44.1	73.4
S. boulardii	116	48.6	20	45.2	66.5

sons : never received study drug or missed more than 3 doses (26), diarrhea within 24 h of starting study (15), less than 72 h of antibiotic therapy (12), exclusion drug started (9), or radiation therapy started (2). There were no significant differences in demographic characteristics between the evaluable and nonevaluable patients nor between the placebo and the yeast treated groups (Table 1).

In Fig. 1, the percentages of diarrhea in the study population and in subsets of the study population given *S. boulardii* or placebo are presented. Of the 254 patients included in the study (all patients except the 64 dropped for reasons discussed above), 6.5 % experienced diarrhea in the yeast group and 16.5 % in the placebo group. A more appropriate group to use for diarrhea incidence was obtained by analyzing the 180 patients who were followed for 8 or more days. All 25 diarrhea cases were found in this subset with 14 cases in the placebo (21.9 %) and 11 cases in the yeast group (9.5 %). The yeast also reduced diarrhea in *C. difficile* positive patients from 31.3 % to 9.4 %. These symptomatic patients would in most settings be considered to have had *C. difficile* associated diarrhea because of the presence in stool of *C. difficile*. The results of Fig. 1 show that, depending on the population analyzed, a consistent 2-3 fold reduction of diarrhea was achieved in patients given *S. boulardii*.

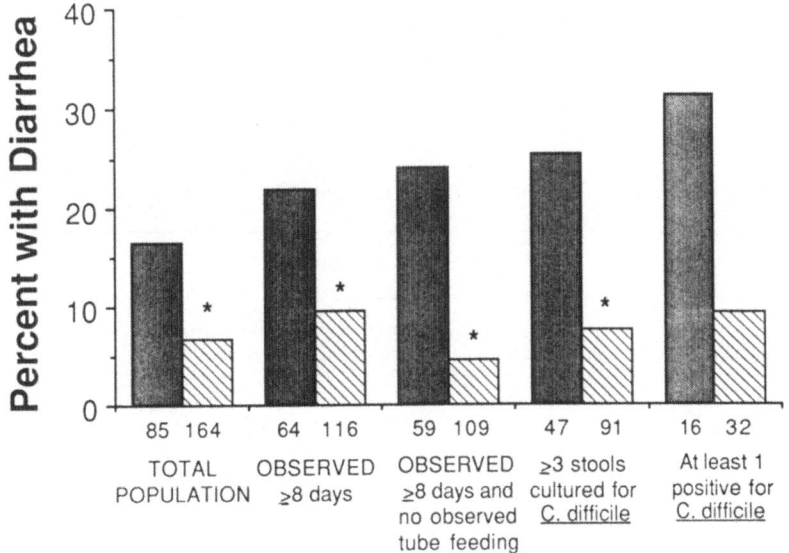

Fig. 1. Diarrhea in the study populations. *Solid bars* : placebo patients ; *hatched bars* : patients on *S. boulardii*. The *asterisk* indicates a significant difference ($p < 0,05$). The *numbers* at the bottom of the bars indicate the size of the population at risk

Antibiotic regimens were analyzed for their ability to increase the risk for diarrhea by a procedure that took into consideration the days of exposure of specific antibiotics (incidence density calculation). The risks were calculated relative to the 31 patients receiving only penicillins. Only 2/21 developed diarrhea and penicillin G was used in both. No single antibiotic used alone showed a significant increased risk for diarrhea compared to penicillins nor did nonspecific multiple antibiotic combinations. However, multiple antibiotic combinations containing clindamycin, or cephalosporins, or ß-lactams, or trimethoprim/sulfamethoxazole increased the diarrhea risk (Table 2). A review of the diet status of patients revealed an increased risk for diarrhea in patients who received antibiotics and were receiving enteral (tube) feeding (Table 2). If these tube feeding patients were eliminated from the analysis, the rate of diarrhea in the *S. boulardii* group was 4.6 % compared to 22 % for the placebo (Fig. 1).

The characteristics of the patients who were *C. difficile* positive are compared to those that were *C. difficile* negative in Table 3. Only patients who had 3 or more stools analyzed for *C. difficile* were included in the calculation of these data. The only significant difference between the two groups was the longer average residence time in the hospital for the patients who were *C. difficile* positive. *S. boulardii* treatment did not affect the *C. difficile* carrier state because the percent of positive patients was about the same in treated or placebo group.

There was no significant association of the presence of *C. difficile* or toxin B in stool with diarrhea (Table 4). It should be noted, however, that the viable counts of *C. difficile* and the titers of toxin B were relatively low, and no patients had overt signs of colitis.

Table 2. Risk factors for antibiotic associated diarrhea*

Factor	Number		Rel risk**	CI
	Diarrhea N = 25	No diarrhea N = 155		
Antibiotics				
Penicillins (single agent)	2	29	1.0	
All multiple	18	87	1.8	NS
All multiple containing :				
Clindamycin	7	8	11.0	3.2,38
TMP-SMZ	4	5	8.6	2.1,35
Cephalosp	8	53	2.9	2.7,4
ß-lactam	17	82	4.1	1.1,16
Tube Feeding	5***	5	4.3	1.8,10

* Calculated from days of antibiotic exposure prior to onset of diarrhea
** Compared to penicillin alone
*** The two cases where diarrhea occured before tube feeding were not counted

Table 3. Characteristics of *C. difficile* positive patients*

Characteristic	Number		p <
	C. difficile pos (n = 48)	*C. difficile neg* (N = 90)	
Days in hospital	21 ± 17	13 ± 12	.002
Age	47 ± 21	45 ± 17	ns
Sex (% male)	34 (71)	61 (68)	ns
Diarrhea	8 (17)	11 (12)	ns
Tube feeding	3 (6)	9 (10)	ns
S. boulardii	32 (67)	59 (66)	ns

* Only patients having 3 or more stools analyzed for *C. difficile* were included ; the numbers in parentheses are the % positive

Table 4. *C. difficile* and antibiotic associated diarrhea

Stool analysis		Number				
		Diarrhea			*No diarrhea*	
	pos	*neg*	*%*	*pos*	*neg*	*%*
C. difficile culture	8[a]	11	42	40[b]	119	34
Toxin B	3[c]	5	38	14[d]	32	44
C. difficile status on entry	4	3[e]	57	11	21[e]	34

[a] Mean viable count = 5.8 ± 1.2 per g stool
[b] Mean viable count = 5.0 ± 1.4 per g stool
[c] Toxin titer = 1.7 ± 0.9
[d] Toxin titer = 1.5 ± 1.3
[e] Negative on entry but positive before exit

Discussion

To our knowlege, the present study was the first to evaluate, in a double blind study, a yeast to reduce antibiotic-associated diarrhea in an acute care setting. A significant prophylactic effect of *S. boulardii* was observed with no apparent adverse effects. The antibiotic associated diarrhea incidence observed in our study was substantial. The placebo population experienced a 22 % diarrhea incidence, a value in general agreement with other reports of antibiotic associated diarrhea frequency. The fact that the average number of days of diarrhea in all patients was 4.5 ± 2.9 with a range of 2 to 11 days points to the persistence of the problem.

Although *S. boulardii* has been used for some time, the mechanism of its action against diarrhea is unknown. The yeast is known to have antagonistic activity against several pathogens in vitro and in vivo [2, 7, 8, 9, 16, 20]. *S. boulardii* has also been reported to increase the disaccharidase activity of the intestinal mucosa [4] and to increase the production of secretory IgA [5]. A better understanding of the mechanism of *S. boulardii* is needed and this knowledge may be useful in devising other ways to favorably modulate the intestinal flora to prevent diarrhea.

Patients at high risk for antibiotic-associated diarrhea were found to be those given multiple antibiotics containing clindamycin or cephalosporins or ß-lactams or trimethoprim-sulfamethoxazole. It is these patients who would presumably benefit most from *S. boulardii* prophylaxis.

Studies in animal models of pseudomembranous colitis had shown *S. boulardii* treatment to be protective [7, 9, 16, 20]. We therefore were interested in the influence of *S. boulardii* on *C. difficile* associated diarrhea in this study. Our results, however, indicated that *C. difficile* played a relatively minor role in antibiotic-associated diarrhea at Harborview Medical Center at the time of the trial. A large number of asymptomatic *C. difficile* carriers were enrolled in this study (33 % of the nondiarrhea group about the same as the number of positive patients in the diarrhea group) and nearly one half of these culture positives were cytotoxin positive. These data imply that the finding of *C. difficile* and cytotoxin in a diarrheal stool does not necessarily implicate *C. difficile* as the etiological agent. While *S. boulardii* did not influence the number of patients who were or became *C. difficile* positive while in the hospital, treatment with the yeast did reduce diarrhea in the *C. difficile* patients. Whatever the etiology of antibiotic-associated diarrhea, the beneficial effects of *S. boulardii* in decreasing this adverse consequence of antibiotic therapy were evident in our study.

References

1. Adam J, Barret A, Barret-Bellet C (1977) Essais cliniques contrôlés en double insu de l'ultra-levure lyophilisé. Étude multicentrique par 25 médecins de 388 cas. Gaz Med Fr 84 : 2072-2078
2. Brugier S, Patte F (1975) Antagonisme in vitro entre l'ultra-levure et différent germes bactériens. Med Paris 45 : 3-8
3. Bowden TA, Mansberger AR, Lykins LE (1978) Pseudomenbranous enterocolitis : mechanism of restoring floral homeostasis. Am Surg 74 : 178-183
4. Buts JP, Bernasconi P, Craynest M, Maldague P, DeMeyer R (1986) Response of human and rat small intestinal mucosa to oral administration *Saccharomyces boulardii*. Pediatr Res 20 : 192-196
5. Buts JP, Bernasconi P (1987) Stimulation of secretory Iga secretory component of immunoglobulins in the small intestine of rats treated with *Saccharomyces boulardii*. IX International symposium on Gnotobiology, Versailles, France, June 21-26
6. Colombel JF, Cortot A, Neut C, Romond C (1987) Yogurt with *Bifidobacterium langum* reduces erythromycin-induced gastrointestinal effects. Lancet II : 43

7. Corthier G, Dubos F, Ducluzeau R (1986) Prevention of *C. difficile* induced mortality in gnotobiotic mice by *Saccharomyces boulardii*. Can J Microbiol 32 : 894-896

8. Ducluzeau R, Bensaada M (1982) Effect comparé de l'administration unique ou en continu de *Saccharomyces boulardii* sur l'établissement de diverses souches de *Candida* dans le tractus digestif de souris gnotoxenique. Ann Microbiol 133B : 491-501

9. Elmer GW, McFarland LV (1987) *Saccharomyces boulardii* suppression of overgrowth of toxigenic *Clostridium difficile* following vancomycin treatment in the hamster. Antimicrob Agents Chemother 31 : 129-131

10. George WL, Sutter VL, Citron D, Finegold SM (1979) Selective and differential medium for isolation of *C. difficile*. J Clin Microbiol 9 : 217-219

11. George WL, Rolfe RD, Finegold SM (1982) *Clostridium difficile* and its cytotoxin in feces of patients with antimicrobial agent-associated diarrhea and miscellaneous conditions. J Clin Microbiol 15 : 1049-1053

12. Gotz V, Romankiewicz JA, Moss J, Murray HW (1979) Prophylaxis against ampicillin associated diarrhea with a lactobacillus preparation. Am J Hosp Pharm 36 : 754-757

13. Gorbach SL, Chang T, Goldin B (1987) Successful treatment of relapsing *C. difficile* Colitis with *Lactobacillus* GG. Lancet II : 1519

14. Kleinbaum DG, Kupper LL, Morgenstern H (1982) Epidemiologic research principles and quantitative methods. Lifetime Learning Publications. Belmont CA, USA

15. McFarland LV, Coyle MB, Kremer WH, Stamm WE (1987) Rectal swab cultures for *C. difficile* surveillance studies. J Clin Microbiol 25 : 2241-2242

16. Massot J, Sanchez O, Couchy R, Astoin J, Parodi AL (1984) Bacteriopharmocological activity of *Saccharomyces boulardii* in clindamycin induced colitis in the hamster. Arzneim Forsch 34 : 794-797

17. Rao SSC, Edwards CD, Austen CJ, Bruce C, Read NW (1988) Impaired colonic fermentation of carbohydrate after ampicillin. Gastroenterology 94 : 928-932

18. Rolfe RD (1984) Role of volatile fatty acids in colonization resistance to *C. difficile*. Infect Immun 45 : 185-191

19. Schwan A, Sjolin S, Trottestam U (1983) Relapsing *Clostridium difficile* enterocolitis cured by rectal infusion of homologue feces. Lancet II : 845

20. Toothaker RD, Elmer GW (1984) Prevention of clindamycin — induced mortality in hamsters by *Saccharomyces boulardii*. Antimicrob Agents Chemother 26 : 552-556

21. Van der Waaij D, Bergkuis JM, Lekkerkert JEC (1972) Colonization resistance of the digestive tract of mice during systemic antibiotic treatment. J Hyg 70 : 605-610

22. Van der Waaij D (1982) Colonization resistance of the digestive tract ; clinical consequences and implications. J Antimicrob Chemother 70 : 263-279

Achevé d'imprimer par Corlet, Imprimeur, S.A.
14110 Condé-sur-Noireau (France)
N° d'Éditeur : 233 - N° d'Imprimeur : 10975 - Dépôt légal : avril 1989
Imprimé en C.E.E.